Contents

Preface

In the early 1990s, Ethiopia entered a period of exhaustion. After 30 years of military conflict and periodic famine, the fighting was suspended and good rains brought forth good harvests. There was a change of government and secession of a northern province, both of which passed off with remarkably little violence. Some people hailed this new era as Peace; others called it a lull between storms.

This book argues that it may well be both. Peace is a precondition for life without famine. The loss of human and economic resources associated with war is too great to permit the scale of national reconstruction necessary for adequate famine prevention. Peace therefore should be hailed as the beginning of the end of famine. Yet the absence of war is not sufficient to prevent starvation. The latter rests on eradicating food and personal insecurity among vulnerable households. This takes time and money. Neither may be available in sufficient quantity to prevent another famine from occurring within the next 10 years, but with the resources and information that are available, we must try.

The data collection upon which this book is based was conducted in 1989/90 when the fighting in Ethiopia was coming to a climax, but in a lull between famines. It was an appropriate time for taking stock of the past and planning for the future. Between 1988 and 1993, numerous government ministries, donor agencies, and nongovernmental organizations collaborated in this research. As a result, we have incurred many professional debts during the period of work.

We are indebted to the German Agency for Technical Cooperation (GTZ) for the funding that made the research possible and to the following international organizations for their cooperation during the project: the International Livestock Centre for Africa, the Food and Agriculture Organization of the United Nations, World Food Programme, United Nations Emergency Preparedness and Prevention Group, UNICEF, EEC, Redd Barna-Ethiopia, Save the Children Fund (UK), Lutheran World Federation, OXFAM America, and CARE-Ethiopia. Government collaboration was provided by the Ministry of Planning and Economic Development, the Ministry of Agriculture, the Relief and Rehabilitation Commission, and Addis Ababa University.

The following individuals served as thoughtful discussion partners during our work of the past 5 years: Judith Appleton, Sarah Atkinson, Shewangizaw

Bekele, Jim Borton, Ray Brokken, Jane Brown, Eshetu Chole, D. Layne
Coppock, Simeon Ehui, Gary Eilerts, Graham Farmer, Hank Fitzhugh, Step-
hen Franzel, Aklu Girgre, Michael Goe, Tsehaye Haile, Michael Harvey,
Wolfgang Herbinger, Sarah Holden, Marion Kelly, Randolph Kent, Allen
Jones, Alemayu Lirenso, Ingo Loerbroks, Charles May, Karen Moore, David
Morton, Richard Pankhurst, Dessalegn Rahmato, Thomas Reardon, William
Renisson, Frank Riely, Jr., Stephen Sandford, Senait Seyoum, Gary Smith,
Tesfaye Teklu, Cornelius Tuinenberg, Michel Vallee, Steven Vosti, John
Walsh, Timothy Williams, Amde Wondafrash, and Eshetu Zerihun.

A large vote of thanks goes to Yohannes Habtu, Luel Getachew, and Raya
Abagodu for their excellent research assistance in Ethiopia, without which
this book would not have been possible. Thanks also to the enumerators:
Abrahim Abdulai, Abdulsalaam Aliye, Jemal Beker, Bazabe Bonga, Debasso
Elema, Jemal Aba Fogi, Araya Gebre-Medhin, Chanyalew Gebre-Sellasie,
Danyachew Gebre-Sellasie, Molle Gelgelo, Tadesse Genebu, Fikre Sellasie
Getachew, Shanu Godana, Abayneh Haile, Tadele Haile, Asrat Mana, Ka-
dhir Mohammed, Yirgalem Negussie, Ahmed Ousman, Hossein Tadele,
Emebet Wolde-Aragay, and Zenebe Yeuhalawork.

In Washington, DC, first class research and technical support was provided
by Yisehac Yohannes. Additional invaluable help came from Ourania Korka,
Lynette Aspillera, and Jay Willis.

The manuscript benefited greatly from thoughtful reviews of earlier drafts
by Melaku Ayalew, Christopher Delgado, Mersie Ejigu, Jonathan Olsson,
Rajul Pandya-Lorch, Edgar Richardson, Peter Svedberg, Sudhir Wanmali,
Mesfin Wolde-Mariam, and Tesfaye Zegeye.

Personal thanks are owed too. Patrick Webb wishes to thank Anna for her
tireless support both in the field and during the writing. Without her much less
would have been possible. Thanks also go to Tristan and Jean-Julien for their
priceless inspiration. And thanks also go to his parents and brother for their
enduring confidence. Finally special thanks go to Jim Hanwell for providing
the original spark and to Per Pinstrup-Andersen for kindling the vision that
continues to motivate work such as this.

Joachim von Braun is grateful to Barbara for her encouragement and cri-
tiques, and to Leonie, Amrei, and Viola for their acceptance of his many and
long absences from home for the purpose of this study.

Finally, we are grateful to the farmers and pastoralists who made us welcome and who shared their pain with us. Many of those who made us welcome
in the late 1980s did not survive the renewed crises of the 1990s. Some of them
will not survive the next famine when it comes. But few Ethiopians give up
hope. It is this unquenchable hope in a better future that compels us to try as
hard as they do to turn hope for a better future into reality.

1 Introduction

This book is about life, not death. It does not dwell on the misery associated with famine. Rather, it seeks to understand how and why some people survive. Although millions of Africans have succumbed in recent years to the disease, starvation, and social dislocation that characterize famine, many millions more have struggled through the crises and lived on to fight another day. The survivors are weakened by the struggle, and many fall prey to the next tragedy down the road. Yet, most carry on in the belief that although things are likely to get bad again in the near future (another drought may strike), life should improve in the long run.

What can we learn from this? Two assumptions about the value of studying famine survivors underpin this book. The first is that we can learn a great deal about chronic and acute food insecurity if we stop generalizing. Some regions of a country are devastated by famine whereas others are not. Some people die in the devastated areas and others do not, even in the same household. Famines act selectively not universally. The journalistic image of famine as the "great leveler", whereby "almost everyone in the affected areas suffers total deprivation," is challenged by just one hour spent in a famine-hit village (FAO 1992). Some people have lost all that they own, some have gained. Therefore the idea that famine is an homogeneous actor affecting an amorphous group of victims must be reconsidered.

The second assumption is that if groups of people are affected by, and respond to, famine in different ways, then relief interventions should be better tailored to location- and income-specific needs of such groups. For example, food aid distribution may be required in one locality whereas cash distribution may be a better option elsewhere. Similarly, poorer farmers may not be able to participate in food-for-work projects owing to a lack of resources to hire labour, so relatively richer farmers may benefit more than others from such projects. In other words, a search for universal solutions to famine is inappropriate. As manifestations of collective failure, famines have complex origins that usually require complex answers. Swift and appropriate action by nongovernmental organizations (NGOs) and governments must therefore be based on a solid understanding of the experiences of the people most involved. In practice, this means the poor and dispossessed and the potentially poor and dispossessed.

It is here that a detailed empirical study of famine among households in Ethiopia has a bearing on actions across Africa. Ethiopia can be seen as a

microcosm of much that is African. It is, for example, the second most popu-
lous country on the continent—roughly one in ten Africans is an Ethiopian
(USAID 1993). The country boasts over 85 ethnic groups bearing allegiance
to most major world and animist religions. Several dozen agro-ecological
zones cover the full range from true desert (the hottest place in the world, in
terms of average annual monthly temperature, is in the Danakil Depression)
to snow-covered mountain peaks to tropical forests. There are more livestock
per person in Ethiopia than in any other part of the continent. In other words,
there are few social, agricultural, or geomorphological conditions in Sub-
Saharan Africa that do not have some manifestation in Ethiopia.

Unfortunately, the same diversity is true of Ethiopia's political and economic conditions. The civil and cross-border wars, political experimentation,
and economic decline found in much of Africa during the past two decades
have all been represented in Ethiopia. Abyssinia's feudal society was shat-
tered by revolution in 1974, giving way to 17 years of communist-style total-
itarianism. The post-1974 regime survived an armed invasion by Somalia in
1977, largely through support from Soviet allies. At the end of the Cold War,
however, such support disappeared. In 1991, after years of bitter fighting, the
regime was defeated by a coalition of opposing guerilla forces marshalled by
the Tigrai-dominated Ethiopian Peoples' Revolutionary Democratic Front
(EPRDF). This opened the way to a transitional government which committed itself to laying the groundwork for the first free national elections in
Ethiopia's history.

At the same time, the 30-year war between Ethiopia's central authorities
and the region of Eritrea came to a close when Ethiopia recognized Eritrea's
independence on May 3, 1993. Large-scale war, it seemed, was a thing of the
past and the role of conflict in famine could now be discounted. Predictions of
9 percent growth in agricultural output for 1993/94 underscored the relative
optimism of the new post-war Ethiopia (USAID 1993).

FAMINE AS A DEVELOPMENT ISSUE

Despite its semblance of calm, Ethiopia, like much of Africa, has little more
than limped through to the twenty-first century under a heavy burden of
human and economic problems. These burdens include continuing misery
among millions of households, characterized by high levels of malnutrition
and food insecurity, crippling debts (accumulated at both national and household levels), a natural resource base facing severe degradation, and rates of
population growth still exceeding real economic growth.

Much has been done to address these chronic problems. Government policies in the 1990s developed rapidly in favour of market liberalization, the
dismantling of parastatal organizations, and disengagement of the public sector from economic management. Steps were taken to improve democratiza-

tion and the rule of law. The international finance community became willing, for the first time in 20 years, to invest development capital in the country as well as supporting a "social safety-net" for softening the blow of macroeconomic adjustment. These measures, coupled with a reallocation of resources away from military spending towards rural development, were all essential first steps in the eradication of famine.

But, are such steps sufficient in themselves to prevent another famine? It can be argued that with an end to war and the foundations for longer term economic growth in place, poor countries will ultimately grow themselves out of vulnerability. The market, freed from its former straight-jacket, will provide the food (either through domestic production or through imports) and income necessary to meet minimum requirements. Famine, as represented by the disruption of society through the cumulative failure of production, distribution, and consumption systems, will disappear. Under this scenario, studying famine is like studying the last specimen of an endangered species.

There are two main reasons why this is unlikely to be true in the near future. First, vulnerable households are not great competitors in the marketplace. The 10 years from 1983 to 1993 witnessed three major droughts, one national famine, the escalation and culmination of a civil war, experiments in the wholesale relocation of villagers, drastic reversals in government policy and structural adjustment. These numerous upheavals have contributed to a depletion of households' assets and savings, declining labour productivity linked to the erosion of natural and human resources, and a continued reluctance among farmers to invest in land productivity enhancement. This process has left millions of Ethiopians worse off than they were at the start of the 1980s. It was estimated in 1992 that between 5 and 15 million individuals would be vulnerable to famine if economic conditions got any worse (Habte-Wold and Maxwell 1992; World Bank 1992; Ethiopia-TG 1993).

These individuals (representing more than 10 percent of the total population) are unable to respond to improved market incentives in the short term. There are two main reasons for this. First, they do not have the resources (or credit) to grow more food even if they wanted to sell a surplus on the market tomorrow. Oxen, seeds, hired labour, and tools all require capital, which is in short supply in rural areas. Second, the transport and marketing infrastructures that are required for the smooth flow of food, capital, and labour around the country are still absent. The costs of rapidly increasing food production are rising almost as fast as the heralded benefits. Consequently, even with growth apparent in the overall economy, the chronically poor will remain vulnerable to climatic or economic shocks for many years to come.

The second argument in favour of continued analysis of famine is that the problem itself is changing. The ground is shifting beneath our feet. Structural adjustment, the freeing of labour markets, rapid urbanization and growing urban food insecurity, the demobilization of huge armies, regional fragmentation of decision making attendant on democratization, all of these processes cast the famine problem in a new light. There are "new" groups of vulnerable people to be considered, new structural problems to be addressed by policy and project action, and new roles for institutions to be considered. And, with a population growth rate of around 3 percent per year, the numbers of people involved are burgeoning. Ethiopia is expected to have a population of over 70 million people at the turn of the century (Ethiopia-CSA 1991). Even if the relative share of vulnerable people does not change, the absolute number of individuals requiring assistance is becoming hard to manage.

Thus, famine is a development issue and not just a humanitarian concern. The artificial wall between relief and development must be breached. Rebuilding the food security of millions of vulnerable households therefore has to be a central pillar of development in any famine-prone country.

For this, there is no silver bullet; no easy or inexpensive solution. There are finite resources in the donor community and among poor governments for investment against famine. However, many resources used in Africa (be they food, technology, or human skills) are fungible; that is, they can be interchanged for a variety of purposes. For example, food aid or the skills of voluntary nurses should not be mobilized only when feeding camps and aid distribution centres are set up to counter a disaster. The feeding camp is a symbol of failure; it signals that policymakers failed to invest their resources more appropriately in the recent past.

Policymakers and planners, awakening late to the realization that *not* preparing against famine can be more expensive in the long run than responding to individual crises, are beginning to ask the right questions about more appropriate resource allocations. But, without information on the impact and costs of alternative policies and projects, few sound decisions can be taken about investment alternatives. The longer that decisions are made on an uninformed *ad hoc* basis, the longer the threat of famine will persist. The longer famine persists, the greater the problem becomes.

AIMS OF THIS STUDY

This book does not attempt to develop new theories or models of famine. It has the more modest goal of providing empirical evidence about the experiences of rural households during and after famine, and of suggesting what might be done to assist those households. Although a mountain of literature about famine exists, our understanding of the reality of famine has been hampered for years by a lack of detailed information about conditions on the ground. Given the large amounts of money and energy spent in saving people's lives in Africa, we still know surprisingly little about those lives.

It is obvious that the pot-bellied individuals portrayed on harrowing fundraising posters are in a desperate condition. We know that they have probably sold or mortgaged all of their possessions, that they have no savings left, that even if rescued from starvation they may not make it through the next crisis. We know, almost subconsciously, that the process leading to this tragedy was dehumanizing. But little is known, even by the NGO staff that save lives, about why things became so bad for those individuals being helped.

For example, Torry (1984) argues that, "quantitative household surveys documenting crisis-induced losses, sacrifices, and adjustments . . . are scarce." Goyder and Goyder (1988) note that there remains "a number of serious gaps in our understanding about how Ethiopian famines affect individual households and how they cope with famine." Cutler (1985) points out that "research into human response to drought should be an urgent undertaking. For Ethiopia, in particular, we need to know why some populations are more vulnerable than others." Similarly, Campbell (1990) argues that "while most studies have described the responses of particular societies to specific crises, they have not analyzed differences within and between groups in ability to cope." And Wolde-Mariam (1991) affirms that "no attempt has so far been made to study disparities between highlands and lowlands, or between the different parts of the highlands."

This study addresses some of these information gaps in relation to the experiences of famine at the grassroots level. It focuses on the details of day-to-day activity that characterize human resilience in the face of desperate odds, disaggregating data across different income groups and across different agro-ecological zones. Four key questions are addressed:

- What types of households are worst affected by famine?
- What do these groups do themselves (through "coping mechanisms") to minimize the effects of famine?
- What importance should be ascribed to factors such as drought, war, and market failures in explaining famines?
- Are the most vulnerable households effectively reached by famine relief and rehabilitation projects? For example, how effective are food aid, feeding camps, employment-generation activities (public works), asset-distribution projects, and agricultural-technology transfers in mitigating famine and in rebuilding the post-famine economy?

Although the emphasis throughout is on a household level analysis specific to Ethiopia, reference is made where appropriate to examples drawn from other parts of famine-prone Africa. Whereas the local manifestations of famine vary, many of the underlying conditions relevant to famine causation and evolution are common across many regions and countries: namely, proneness to production fluctuations, lack of employment opportunities, limited household assets, isolation from major markets, low levels of farm technology, constraints to improvements in human capital, and poor health and sanitation environments.

Thus, although only broad generalizations can be drawn from the minutiae of individual case studies, the detailed information gained may lead to a reassessment of generalizations that are accepted as conventional wisdom. At the same time, microlevel analysis must be set into the wider context of regional and national level processes. This is particularly important where discussions of famine causation come into play. The roles played by drought, military conflict, and market collapse are all examined at this higher aggregate level.

DATA SOURCES AND METHODOLOGY

The analysis is based both on fieldwork and on published data sources. The published data derive largely from Ethiopian government agencies, such as the Central Statistics Authority, the National Meteorological Services Authority, the Ministry of Agriculture, and the former Office of the National Committee for Central Planning (now the Ministry of Planning and Economic Development). Although much maligned, government figures were rarely found to be overtly manipulated. Indeed, it can be argued that Ethiopia possesses a greater wealth of usable published data than most other countries in Africa.

However, it should be noted that the quality and quantity of such data often leave something to be desired. Wherever possible, cross-checks were made between government statistics and those of international donor or nongovernmental organizations; where figures were found to be irreconcilable, they were discarded. Nevertheless, care should always be taken in interpreting secondary data. A special note is made in the text and in the footnotes to tables where particular caution is required.

The primary data represent the fruit of intensive, repeat-visit interviews among 550 households in seven different parts of the country. The interviews were conducted from February 1988 to January 1989 by a team of 22 carefully trained Ethiopian enumerators (male and female), closely and continuously supervised by three Ethiopian supervisors and ourselves. Each enumerator was educated to secondary school level and spoke at least two local languages. Most had previous experience of quantitative survey techniques, but were given comprehensive retraining prior to the commencement of data collection.

Given the distressing nature of the interview topics, great care was taken to get to know the respondents, to spend as much time as possible with them, and to be sensitive during interviews. The enumerators (divided into three groups) spent up to three months at each survey locale. Individual households were visited between three and seven times during this period, and both men and women participated in the interview process (in separate sessions). The multiple visits permitted collection of a broad array of data from each household, as well as on-site cross-checking and verification of information gathered.

PROBLEMS IN THE STUDY OF FAMINE

A number of limitations of the book should be openly declared at the outset. Conducting research on famine is not easy. Spending resources on surveys at the height of a crisis, instead of using them for relief, is intrusive and possibly unethical. The information obtained at such a time may also be so full of biases (respondents altering facts according to their expectations of assistance) as to be unusable. For these reasons, there are exceptionally few studies of actual famine conditions other than those dealing with nutrition and health issues (Appleton 1987; Rivers 1988).

The post-famine environment is only slightly more favourable. The real losers, those who died or were forced to move away, are no longer available for interview. This biases any sample frame in favour of households that survived. Of course, famine survivors are themselves under grinding stress for long periods after a crisis. It is hard for people to talk about the traumas of the recent past. Yet, recent recall of famine experiences among the people still present is a first best alternative to actual measurement during the crisis.

This was the approach used for the present study. Individuals were carefully asked, through structured questionnaires and open-ended discussions, to explain what happened to them in the preceding four years and why. No illusions are entertained about the completeness of the data obtained in this way. Memory is always selective. However, the interview of several people within a household allowed for the confirmation or refutation of uncertain responses.

The second limitation relates to survey site selection. No attempt was made in the selection of sites to ensure "representativeness" at a national level— only a full-scale national sample survey could aspire to such a standard. The survey locations used in this study were chosen for their positive deviance. That is, after ten months spent in Ethiopia consulting reports and discussing with government and donor officials, only ten sites were found to meet three desired conditions: first, proneness to extreme fluctuations in food production and documented indications of recent food crisis; second, diversity of agro-ecological conditions; and third, public relief or rehabilitation interventions for which adequate baseline information was accessible (thereby permitting an assessment of effectiveness based on original relief objectives and an understanding of implementation constraints).

Although a number of the potential sites were identified in Wollo and Tigrai, the heartland of the 1984/85 famine, military conflict in those areas precluded their selection. Insecurity in parts of Gojjam and the Shewa/Wollo border, far from the actual fighting, also ruled out such areas. The final seven locations were therefore located in territory that was administered by the Mengistu Government.

This did not necessarily make the surveys much easier. For example, the survey sites in north Shewa were under constant threat of guerilla action.

Remote sites in the south of the country, located 60 miles from the nearest provincial town along unmarked dirt tracks, were prone to banditry. Sites in the central regions were reached by unbridged rivers that could be crossed only after a temporary ford was constructed out of stones piled into the water. After heavy rains, such rivers were unfordable for days.

The third limitation relates to substance. More emphasis is placed in this book on the microeconomics and anthropology of famine than on a comprehensive analysis of the political economy of famine. There are political trade-offs in the allocation of investment towards famine prevention, as well as competing and conflicting goals among donors and ministries and power seeking by actors in both the private and public sectors (Pinstrup-Andersen 1993). These interactions among key agents of change are complex enough in liberal democracies. However, the complexity is compounded in a dictatorship where the major players, pursuing rational goals that reflect self-interest, are further constrained by the knowledge that wrong actions can lead to their demise.

The *realpolitik* of Mengistu's Ethiopia is clouded by a lack of documentary evidence about past decision-making processes, and by the too obvious biases of more recent commentators (Clay and Holcomb 1985; Lemma 1985; Wolde-Giorgis 1989; de Waal 1991). There is no doubt that the war with Eritrea, forced resettlement, villagization, the quota system, and the nationalization of land, the litany of woes commonly raised to explain Ethiopia's famines, were all contributing factors. We address some of these factors in this book. However, a detailed analysis of the political economy of famine, beyond the scope of this book, is required urgently for a fuller understanding of how and why policies and resource allocations are made in the way that they are.

STRUCTURE OF THE BOOK

The analysis begins in Chapter 2 with an exploration of what famines are and why they occur. The definition of famine is discussed in relation to the concept of food security, a term which encompasses food supply and food demand factors. The question of causation is considered in relation to drought-induced production shortfalls, agricultural policies, the role of market success and failure, and the problem of war.

Providing an overview of major historical and contemporary famines in Ethiopia and elsewhere in Africa, the third chapter considers factors that are common to famine-prone countries. It briefly examines the record of drought sequences of the 1970s and 1980s, highlighting the correlation with crises on the ground.

Chapter 4 considers the household-level effects of famine. This chapter examines the impact of drought on the production, income, and consumption of smallholder farmers, as well as their responses to that impact. Patterns of household response to drought-associated famines are grouped into three stages—risk minimization, risk absorption, and taking risks to survive. The analysis is pursued at three levels of disaggregation: by income group, by upland/lowland agro-ecological zone, and by gender of the head of household.

In Chapter 5, the household-level analysis of production and consumption issues focuses on the special case of pastoral communities. Given their relatively greater social cohesion and mobility, the impact of drought and famine on pastoralists is very different to that on farmers. However, the long-term vulnerability of pastoralists may soon exceed that of settled farmers, given new pressures on "common" land due to the recent revisions in land tenure policy, removal of restrictions on labour mobility, and population migration out of densely inhabited highland regions down to the semi-arid pastoral fringe. This brings the future ability of pastoralists to respond as flexibly to drought as in the past into doubt.

An assessment is made in Chapter 6 of the experience of relief and rehabilitation interventions. Seven different project activities are examined in detail to ask, who participated, why, and what was the outcome? This analysis provides a preliminary step towards an evaluation of alternative policy measures in the fields of agriculture, trade, and infrastructure investment in response to drought.

The final chapter discusses some of the lessons that may be drawn from the analysis and makes a number of recommendations concerning further steps to be taken for improved famine mitigation in Africa as well as prevention. The chapter argues that scarce resources for investment should be concentrated in a relatively narrow and well-targeted range of interventions. Given that continued famine in Africa represents a failure of policy and project design and implementation, a consistent set of policies designed to support improved public interventions is required over the next decade in order to remove Africa's obstacles to rural growth. If the design of such policies and programmes is based on a better understanding of the dynamics of the rural economy, it is argued that famine can indeed be defeated in the next decade.

2   When Plenty is Not Enough

Famine is the product of human hands. Individual human failure, such as the break-up of a household through starvation or migration, has its roots in collective human failure. The collapse of societal norms that characterizes famine is the outcome of a lack of appropriate action. Countless environmental, economic and historical factors can be mustered to explain why a famine takes place at a given time and location. But, responsibility for its occurrence cannot be evaded by those who could have, in most cases, prevented it. This includes community leaders and donor agencies as well as the national governments most involved.

Starting from this understanding, the present chapter does not engage in arcane discussions about terminology. Rather, it spells out the key characteristics of famine in Africa and how these relate to concepts of food security, and moves on to elaborate on the principal manifestations of famine in Africa during recent decades.

DEFINITIONS AND CONCEPTUAL FRAMEWORK

Famines are about death; that much is obvious to the casual observer. But, in reality famine is about more than death. To the author of Revelations in the Bible, Famine is an entity distinct from the other three Horsemen of the Apocalypse—War, Pestilence, and Death. In the latter distinction, famine was seen as a catastrophe separate from war and pestilence and apparently also separate from death. This is the argument followed more recently by Wolde-Mariam (1984) and de Waal (1987) who attempted to disentangle the more synthetic definition by specifying differences between famines that involve "excess" mortality, and those that do not.

Famine is defined here as a catastrophic disruption of society as manifested in a cumulative failure of production, distribution, and consumption systems. According to this definition famine has three principal manifestations.

(1) Extreme, geographically concentrated shortfalls in food consumption that result in chronic loss of body weight and a rise in excess mortality (a net increase above average rates).
(2) Massive social disruption, including community dislocation (increased

distress migration and out-migration of entire families), and "abnormal" behaviour (increased reliance on foraged foods, conflict among neighbours, increased begging).
(3) Long-term resource depletion, including the degradation of productive material assets, of the natural resource base, and of human capital.

Thus, the human death toll, a tragic erosion of human resources, is only one part of the overall problem. A long-term depletion of resources carries the implications of famine far beyond the realm of a discrete event. Once a localized crisis has passed, the process of nation-building is severely impeded by the loss of people, community integration, livestock, savings and even the government's capacity to tax and invest. Such losses make the process of social and economic rehabilitation very difficult because they compound the poverty that pre-dated, and contributed directly to, the individual crisis. The famine survivors of today can easily become the famine fodder of tomorrow.

But, which ones are most at risk? Numerous concepts have been brought to bear on the issue of why some people are more at risk of dying, or moving, or losing than others. Such concepts have brought along with them a veritable lexicon of new buzz-words, such as entitlement failure (Sen 1981; Drèze and Sen 1989), stages of coping (Jodha 1975; Rahmato 1987), exposure, capacity and capability (Chambers 1989), transitory, chronic and acute vulnerability (Maxwell 1989), and empowerment and enfranchisement (Watts and Bohle 1992), to name but a few. These have been used to varying effect, singly or in combination, in attempts to clarify and refine our understanding of what makes the livelihood of one person or region more likely to collapse under sustained or sudden pressure than that of others.

Such understanding evolves by accretion, as new dimensions are added to the debate and the jargon becomes ever more intricate. Quantum leaps in understanding, fueled by new organizing concepts, have been rare since Sen's (1981) work of the early 1980s, who's major contribution was to divert attention away from an emphasis on food production and to focus on the ability of individuals to obtain control over food. In his view, famine results when the command of certain people over basic food necessities, their entitlement, collapses and is not protected by state action (Drèze and Sen 1989). Thus, famine can occur without a significant decline in food supply if there is a significant decline in entitlements among vulnerable groups of the population.

This approach brings analysis to bear on issues such as social and legal access to food commodities, differential access to "entitlement protection" (relief interventions), the terms of trade between commodities as well as between actors in the market place, and the key role of real purchasing power in determining household options during a crisis.

There has been much discussion on the relative merits of the "theory" of entitlements versus other approaches to explaining famine (De Waal 1991; Osmani 1991). Many gaps in its potential application have been identified. For example, Swift (1989) notes that entitlement is an ahistorical explanation that does not treat changing vulnerability over time; that is, the link between past and current or future crises. Watts and Bohle (1992) argue that the entitlement approach is often interpreted too narrowly, thereby failing to account for the social determinants of "power" that account for differential vulnerability across gender, ethnicity, and caste.

Similarly, Locke and Ahmadi-Esfahani (1993) propose that the entitlement focus on legal structures tends to discount the role of illegal transactions (stealing, smuggling). And, Clay (1991) argues that entitlement "theory" is linked inadequately to "the vast body of work written around the organizing concepts of food security". The latter also suggests that the contrast made by Drèze and Sen (1989) between "entitlement protection" (famine relief) and "entitlement promotion" (actions to prevent famine) contributes to the oversimplified dichotomy often drawn between relief and development.

The fact that analysts continue to criticize Sen's work (both constructively and destructively) more than a decade after its publication underlines its seminal contribution to the debate. We do not intend to join in the fray here. However, Clay's question about famine's relation to food security deserves some elaboration.

In many ways, famine is the antithesis of food security. In a simplified sense food security represents the *absence* of conditions necessary for famine. Conversely, food *in*security, an endogenous outcome of resource availability and of policies and potentials dictating resource use, can be seen as one of the roots of famine.

However, the antonym of famine—plenty—is not a sufficient condition for food security. Not all food insecure nations and not all food insecure people are equally, or even necessarily, vulnerable to famine (Habte-Wold and Maxwell 1992; Watts and Bohle 1992). Individuals can die through famine even when food is available. Just as there are a multitude of factors determining whether a drought will trigger a famine, a host of factors determine whether an individual will die during a major crisis (Chambers 1989; Swift 1989). It is a combination of the degree of poverty, the degree of risk of failing to secure local food and income, and the chance of receiving external assistance that defines vulnerability to famine. This is where the concepts of food security and entitlement come together.

Food security can be defined as access by all people at all times to the food required for them to lead a healthy and productive life (von Braun *et al.* 1992). Although food is the defining focus of the concept, food is clearly not all that matters. Food security is an encompassing concept that addresses the risk of individuals and households not being able to secure needed food.

The risk of constrained access to food can arise from many directions. For example, risk of access failure can be driven from the supply side through a collapse in food production. Alternatively it can come from the income side, such as when price explosions occur or terms of trade between key commodities are sharply altered.

Therefore, risk of constrained access to food allows for a consideration of all aspects of livelihood (employment and income patterns) and the environment (physical, health, and sociopolitical), and not just of food consumption and use habits. As such, food security is a concept that crosses the conceptual wall between emergency relief and development activities. The risk of failure in either realm of activity determines vulnerability to an erosion of food security and descent down the slippery slide towards famine.

Figure 2.1 provides a conceptual framework for examining the linkages between famine and food insecurity through various components of the food chain. In this conceptualization, private action in each of the five links of the food chain can, depending on the natural, socioeconomic, and political context of those actions, lead towards improved household food security or towards vulnerability to famine.

For example, the key elements that determine successful food security, food availability, access, and use (in the second row of the diagram), are the outcome of multiple processes of food supply, marketing, and demand operating at both national and household levels. Food availability at the national level is a function of production plus imports (both commercial and food aid), plus existing stocks, minus effective demand. Domestic production rests heavily on the natural resource base of a country or region (box 1, Figure 2.1), as well as on the productive assets (ploughs and tools) and human capital stock (physical capability and education) held by individual households (box 2, Figure 2.1).

Seasonal and interannual fluctuations in production, and therefore in food availability, are serious given the resultant price fluctuations (caused by the dysfunction of markets and institutions), which largely determine accessibility (box 3, Figure 2.1). The other element of accessibility relates to available household purchasing power, reliant on nonfarm income sources and support networks.

Appropriate food use rests on the quantity and quality of food available for consumption (box 4, Figure 2.1), the stability of its availability and access, the distribution of accessed food among members of the household, and the ability of individuals to convert calories that are eaten into the energy required for physical health and activity (box 5, Figure 2.1). Health is important here (often it is forgotten) because adequate income and calorie intake do not always translate into adequate nutritional status. Such is the case if calories gained are lost through diarrhoea, parasites, and episodes of epidemic disease. Thus, health is also a matter of food security.

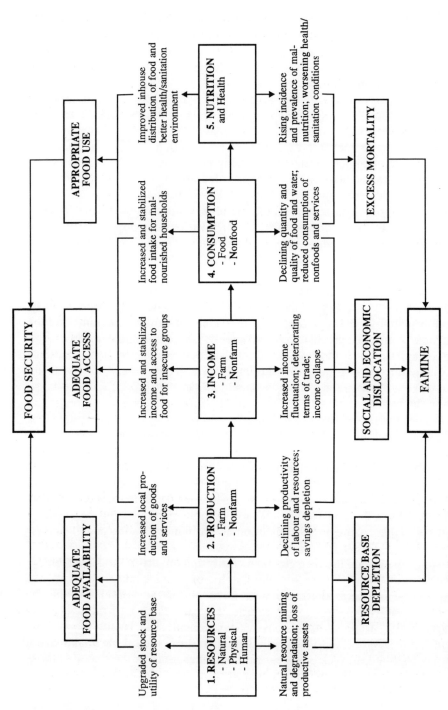

Figure 2.1. A conceptual framework for understanding relationships between famine and food security

By contrast, the major symptoms of famine—resource base depletion, social and economic dislocation (community break-up, market and institutional failure), and human mortality—derive from the failure of many of the processes and events outlined above. A lack of food availability is caused by an inadequate mobilization of natural, material and human resources for food production. When domestic production fails and imports do not compensate for food shortfalls, a mining and draw-down of the existing resource base occurs, thereby impairing future productive potential (boxes 1 and 2, Figure 2.1).

A breakdown in access to available food occurs as markets fail (for policy, logistical, or infrastructural reasons), causing prices to rise sharply and the terms of trade with saleable household commodities (such as livestock, tubers, labour, or services) to collapse (box 3, Figure 2.1). Market collapse is compounded by social collapse as intracommunity and even intrahousehold relations are severed under the stress of deprivation, and as institutional support structures (health services, veterinary services) fail through a diversion of operating funds and personnel.

Under periods of sustained food constraint, the use of food becomes increasingly inappropriate in terms of quantity and quality consumed. The number of meals per day is reduced, the amount of food eaten is reduced, and the types of food eaten deviate increasingly from the norm (box 4, Figure 2.1). The resultant loss of body mass and energy is worsened by a heightened susceptibility to disease (box 5, Figure 2.1).

Public action (government policies and donor interventions) may impact either positively or negatively on private initiatives at any point along the chain. Such action can be in the nature of a long-term investment, such as national soil conservation or primary education programmes (affecting box 1, Figure 2.1) or have a much shorter time horizon, such as emergency health and nutrition rehabilitation of target individuals during a crisis (affecting boxes 4 and 5, Figure 2.1).

The middle ground is occupied by interventions such as food-for-work programmes. The latter increase immediate food access by transferring food resources to needy individuals through the labour market (box 3, Figure 2.1), while simultaneously raising the potential for longer term food availability through road building and watershed management, activities that assist in increasing farm production (box 2, Figure 2.1). In other words, there can be upstream and/or downstream linkages associated with actions taken in any one link of the food chain.

A real-life example may be useful here. Take a food-for-work project designed to build stone terraces (small walls that reduce the flow of soil down a slope), thereby recuperating land that had lain fallow for many years and so raise crop output. The project could have three distinct objectives: namely, to bring an area of land back into production, to raise levels of local production, and to increase food consumption. The success of each goal could be identi-fied and measured separately. For example, 100 hectares of land could be brought back into production (which would count as a success in box 1, Figure 2.1). Maize production might benefit from the reduced erosion and rise from, say, 200 kilogrammes per hectare to 800 per hectare (success in box 2, Figure 2.1), and the food consumption level of participating households might rise from 2000 calories a day per capita to 2500 per day (success in box 5, Figure 2.1).

But, would such a project qualify as a complete success? Not necessarily. If the land brought under cultivation after a fallow period of several decades had been used in the interim by pastoralists for dry season grazing, the new terraces and cultivated fields would deprive them of forage that they had come to rely on every year. The pastoralists could take the case to a regional court of law, but it would likely be proved that the land did previously belong to the farming families who now want to cultivate it again. The pastoralists would be forced away to more marginal lands, where soil erosion could be accelerated. This could be a negative outcome of the project in net terms as judged by box 1 (Figure 2.1).

Another case of distributional imbalance could occur where women con-stitute 80 percent of the actual participants of the food-for-work scheme (doing the work to build the terraces), but only men gain access to the land thus reclaimed as a matter of tradition. It is problem enough that the women would gain only a short-term food security benefit (via the wage income), with men gaining a longer term benefit via their enhanced productivity.

But it can be even worse. Because women would receive training in repair and maintenance of structures generated by the project, the subsequent cultivators of the land (men) could not maintain their assets effectively. Small breaks in the terrace would allow storm water to flood through and erosion might become more serious than it was before the project. This would repres-ent only a qualified success in box 1, coupled with failure in box 3 (Figure 2.1).

Thus, if a public intervention has a negative impact in any one box, or a positive impact in one box blocks transmission of success to other boxes, the net result will have been compromised. The benefits or blockages associated with any intervention can be transmitted upstream or downstream of the intervention itself. The women earning income may contribute in the short term to a measured success in box 4 (Figure 2.1) by raising household food consumption above previous levels. But, if their absence from home due to project participation results in worse child care for infants, poorer sanitation, and hence a lower child nutritional status despite higher calorie consumption, then the real impact of an intervention may be less than it appears at face value.

It is this complexity of processes and potential outcomes that characterizes the study of both food security and famine. The same complexity makes the implementation of policies and projects aimed at either at food security enhancement or famine prevention so difficult. As a result, famine survives in Africa and food insecurity grows apace. The following sections examine the manifestations of both conditions in Ethiopia and elsewhere in Africa.

MANIFESTATIONS OF FAMINE IN AFRICA

Between 1970 and 1990 the number of chronically undernourished people in the developing world as a whole fell from over 940 million to 786 million, representing a fall from 36 percent to 20 percent of the total population of these countries (ACC/SCN 1992). Major strides were made in reducing the global prevalence of malnutrition among infants, children, and mothers. India, for example, achieved a reduction in the prevalence of underweight children of roughly half a percent per year during the 1980s (ACC/SCN 1992). Despite severe economic crises, many Latin American countries, such as Bolivia, Colombia, and Chile, also achieved falling rates of underweight prevalence and infant mortality.

Africa, on the other hand, remained an anomaly. Nutritional trends in most parts of Africa deteriorated during the 1980s. The number of people not eating adequate calories for an active and healthy life rose from 130 million to 170 million during the decade (ACC/SCN 1992). The number of underweight children rose from 20 million in 1975 to 27 million in 1990, and is expected to exceed 30 million by the turn of the century (Pinstrup-Andersen 1993). This upward trend in food insecurity paves the way for future famine.

Once a universal threat to human life, famine is now an African phenomenon. This does not mean that other countries in the world are now immune to famine. Countries undergoing drastic economic and political transition on the shaky foundation of a fragile economy will always be vulnerable to climate- or war-induced food and income shortfalls. Poorly developed markets and institutions may be unable to cope with such shortfalls. Parts of central Asia and Latin America, for example, could in the foreseeable future witness localized famines that raise a need for international intervention. But, Africa is likely to remain the most fertile ground for famine well into the twenty-first century.

The reasons for this are many: proneness to extreme production fluctuations, limited nonfarm employment opportunities, low levels of savings, regional fragmentation of major markets, low levels of farm technology with limited potential for the transferal of Asian style Green Revolution technologies, high rates of natural resource degradation, high levels of illiteracy and school non-attendance, poor health and sanitation environments, generally high population growth rates, high national indebtedness, often poor governance leading to a poor distribution of resources and hence to civil conflict, and the high rates of chronic malnutrition noted above that impede future human potential.

Of course, there are exceptions. Not all of Africa is affected by the same problems, and not everyone in affected countries is prone to famine. Some famines are closely linked to military conflict; others are experienced in countries disrupted by civil or cross-border wars, but are outside the zone of conflict; still others are unrelated to conflict. Figure 2.2 presents a map of Africa identifying countries that experienced famine during the 25-year period from the late 1960s to the early 1990s. Famine has clearly been concentrated in three broad areas of the continent: the Sahel, the Horn, and southern Africa. Some countries other than those identified, such as Zambia, Zaire, Tanzania, parts of Kenya and Burkina Faso, have experienced minor, more localized famines at certain times since the 1960s. However, most Africans exposed to famine in the recent past have inhabited one of a dozen countries.

It is no surprise that all 12 of these worst-affected countries fall among the bottom 30 countries (out of 160) in UNDP's Human Development Index (UNDP 1993). The latter index is computed annually and is based on indicators such as life expectancy, literacy, educational attainment, and real Gross Domestic Product. In other words, famine experience has a close association with a low quality of life.

Estimates of famine-related mortality in these countries are widely divergent, based largely on unreliable sources or guesswork. However, it is generally accepted that the total number of famine deaths since 1968 runs into several million. The bulk of these deaths were associated with major famine events in a handful of countries. For example, the conflict-related famine in Biafra (Nigeria) between 1968 and 1970 is thought to have been responsible for roughly 1 million deaths (OFDA 1991b). Excess mortality between 1983 and 1986 in Ethiopia and Sudan probably exceeded 1 million (Zaman and Parker 1990; Webb, von Braun, and Yohannes 1992). And, as many as half a million may have perished in Somalia during 1992/93 (IHD 1993; Kates 1993). The number of individuals still considered to be vulnerable in the mid-1990s exceeded 30 million in the worst-affected countries (Africa Recovery 1991; FEWS 1993a; Kates 1993).

Although Ethiopia is clearly not unique in its experience of famine, its continued status as one of the most famine-prone countries in the Horn of Africa, coupled with its recent history of drought, war, and political turmoil, make it a good case for more detailed consideration. Roughly 5 million of the 30 million Africans vulnerable to famine in the mid-1990s were located in Ethiopia (Barnhart 1993). Of this 5 million, more than 530 000 were not Ethiopian nationals, but refugees from neighbouring Somalia, Sudan, and Kenya (IHD 1993). Ethiopia has the dubious distinction of being one of Africa's largest receivers as well as senders of refugees; in 1991, an estimated 700 000 Ethiopians were classified as refugees and asylum seekers outside of their own country (IHD 1993). Most of these refugees were escaping the military conflict and threat of renewed famine that continued to hang over the

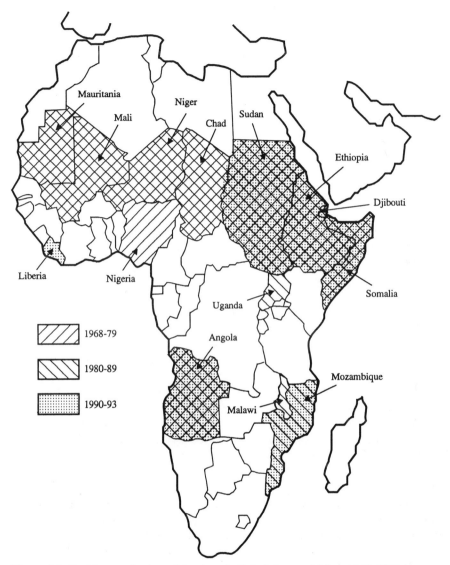

Figure 2.2. Incidence of selected famines in Sub-Saharan Africa, 1968–1993 (sources: Zaman and Parker 1990; OFDA 1991a)

northern regions and parts of the eastern lowlands. The latter constitute some of the areas most prone to chronic food insecurity and famine, still home to at least 4 million highly vulnerable people. In order to better understand the scale and severity of the famine threat in such areas, the following sections examine the geographical and historical background to famine in Ethiopia.

Table 2.1. A chronology of Ethiopian famines and food shortages (sources: Wood 1977; Pankhurst 1984; Wolde-Michael 1985; Wolkeba 1985; Iliffe 1987; Degefu 1987; Gizaw 1988; Gedion 1988; RRC 1990a; FEWS 1991, 1993b; Wolde-Mariam 1991)

Date	Region affected	Attributed causes and severity
253–242 BC	Ethiopia	Deduced from low Nile floods
1066–1072	Ethiopia and Egypt	Deduced from low Nile floods and Egyptian famine
1131–1145	Ethiopia	Severity unrecorded
1252	Ethiopia	First of seven famine years during next 30 years
1258–1259	Ethiopia	Severity unrecorded
1272–1275	Ethiopia	Severity unrecorded
1314–1344	Ethiopia	Severity unrecorded
1435–1436	Ethiopia	Severity unrecorded
1454–1468	Ethiopia	Severity unrecorded
1543–1562	Hararghe	Attributed to God's anger at murder of Emperor Gelawdeos
1618	Northern Ethiopia	Emperor forced to evacuate headquarters
1772–1774	Ethiopia	Widespread human suffering
1796	Northern Ethiopia	Famine triggered by locust invasion
1800	Ethiopia	Large human and livestock death toll
1812–1816	Tigray	Severity unrecorded
1826–1827	Ethiopia	Failure of cotton and grain crops
1828–1829	Shewa	Much human mortality
1831	Tigray	Severity unrecorded
1835–1838	Tigray and Eritrea	Drought, cholera epidemic; high human and cattle loss
1864–1866	Tigray and Gondar	Heavy human death toll
1876–1878	Tigray and Awash Valley	Heavy livestock death tolls
1880	Tigray and Gondar	Much loss of livestock
1888–1892	Ethiopia	Drought and spread of rinderpest caused loss of 90 percent cattle and one-third human population
1895–1896	Ethiopia	Minor drought. Loss of livestock and human lives
1899–1900	Ethiopia	Droughts deduced from levels of Lake Rudolf and low Nile floods
1913–1914	Northern Ethiopia	Lowest Nile floods since 1695. Grain prices said to have risen thirty-fold
1920–1922	Ethiopia	Moderate drought, similar to 1895–1896
1932–1934	Ethiopia	Deduced from low level of Lake Rudolf in Northern Kenya
1953	Tigray and Wollo	Severity unrecorded
1957–1958	Tigray and Wollo	Rain failure in 1957 with locusts and epidemic in 1958
1962–1963	Western Ethiopia	Very severe
1964–1966	Tigray, Wollo	Undocumented. Said to be worse than 1973–1975 droughts

Table 2.1. (*cont.*)

Date	Region affected	Attributed causes and severity
1969	Eritrea	Estimated 1.7 million people suffering food shortage
1971–1975	Ethiopia	Sequence of rain failures. Estimated 1/4 million dead. Fifty percent livestock lost in Tigray and Wollo
1978–1979	Southern Ethiopia	Failure of *belg* rains
1982	Northern Ethiopia	Late *meher* rains
1983–1985	Ethiopia	Sequence of rain failure. Eight million affected. Estimated 1 million dead. Much livestock loss
1987–1988	Ethiopia	Drought of undocumented severity in peripheral regions
1990–1992	Northern, eastern, and southwestern Ethiopia	Rain failure and regional conflicts. Estimated 4 million people suffering food shortage
1993–1994	Tigray, Wollo, Addis	4 million people requiring food assistance, including demobilized army and Somali refugees. New droughts

FAMINE PREVALENCE IN ETHIOPIA

Ethiopia's history is punctuated by famine. Table 2.1 presents a (nonexhaustive) list of periods of severe food shortage and/or excess mortality as recorded in a variety of contemporary and historical sources. Although most of the occurrences listed here fall within the past 200 years (the period for which most detailed records exist), food-related crises can be traced as far back as 250 BC.

Of course, in such a compilation exercise, one faces the problem of defining the events under consideration. Food shortages tend to take on famine proportions over an extended period of time, not overnight. It is therefore difficult to pinpoint the exact start of a famine, and when it ended. Whereas some analysts refer to notable crisis years in Ethiopia, such as 1958 or 1973, others claim that "during the 20-year period between 1958 and 1977, about 20 percent of the country was under famine conditions each year" (Fraser 1988; see also Wolde-Mariam 1984). The analysis depends on the indicators used to define the problem, and on how far back one wishes to extend the chain of causal relationships.

Acknowledging these definitional problems, the list in Table 2.1 identifies 40 periods of crisis, some lasting just a year or two and some apparently persisting for more than a decade. Most events were geographically concentrated into two broad zones of the country. The first comprises the central and northeastern highlands, stretching from northern Shewa through Wollo and

WATERFORD REGIONAL
TECHNICAL COLLEGE
LIBRARY

Class:
Acc. No: 54496

Tigray into eastern Eritrea (Figure 2.3).[1] The second is made up of the crescent of low-lying agro-pastoral lands ranging from Wollo in the north, through Hararghe and Bale (the area known as the Ogaden), to Sidamo and Gamo Gofa in the south. Of the crises listed in Table 2.1, more than half are concentrated in these two zones.

Figure 2.3. Map of Ethiopia indicating the regions most affected by famine. Note that major changes were made in the administrative boundaries in 1989 and again in 1992 (see Figure 2.4). Because the current analysis is based on data collected according to the administrative units that were applicable prior to 1989, most maps in this study are based on the pre-1989 boundaries (sources: compiled from Gebre-Medhin and Vahlquist 1977; Wolde-Mariam 1984; RRC 1985a)

Figure 2.4. Map of Ethiopia's administrative regions proposed in 1992 (source: based on Caldwell 1992)

Why have these two areas been stricken more than others by drought and famine? Although the next chapter pursues the question of famine causation in more detail, it is worth pointing out here that three closely related factors differentiate these two zones from other parts of the country: namely, population pressure, agro-ecological resource base, and climatic regime.

High population concentration in areas of limited natural resources increases vulnerability to food insecurity because the asset base relative to population demands is so shallow. Average population density at a national level in the early 1990s was roughly 40 persons per square kilometre (Gutu, Lambert, and Maxwell 1990). However, more than 80 percent of the population is estimated to be concentrated in highland areas that exceed 1500 metres above sea level, where population densities can reach as much as 300 per square kilometre (Gryseels and Anderson 1983; Hurni 1988).

Upland farmers have migrated in large numbers into the lowlands during the past 100 years as these areas have been progressively cleared for cultivation and malaria eradicated. For example, in northern Shewa hundreds of Amhara farmers left the heavily populated plateau near Ankober during the early 1900s to farm less fertile but more easily obtainable land close to the Rift Valley floor east of Aliyu Amba. Similarly, there was a voluntary movement of villagers during the mid-1980s from the high peaks overlooking the southern town Awassa into the plains bordering the Omo river.

Some migration has also occurred from lowland to upland, and even from mid-altitude uplands to mountain slopes (Wolde-Mariam 1991). Nevertheless, the bulk of the population is still found in the milder, non-malarial lands between 2000 and 3000 metres above sea level, and is located principally in the central and northern highlands (Kloos and Zein 1988; Ethiopia-OPHCC 1991).[2] Given an annual growth rate of more than 3 percent, the national population (excluding Eritrea) is expected to exceed 70 million by the year 2000 (IUCN 1989; Ethiopia-OPHCC 1991). This means that if increasing numbers of people do not leave the most densely populated highland districts and/or resource-enhancing technologies are not widely adopted, pressures on existing natural resources will be aggravated further.

For example, assuming an average need of 0.8 m³ of fuelwood per person per year and current projections for population growth at 3 percent per year, 78 of the 102 pre-1987 administrative areas will face fuelwood deficits by the year 2010. In 20 of the worst-affected areas, meeting the deficit through local reforestation would require the planting of fast-growing eucalyptus on 17% of the total land area—an unlikely scenario (Webb et al. 1994).

In some regions, the concentration of people and animals has already led to deforestation, the cultivation of marginal lands, and a loss of fallow periods for farmed fields (Dejene 1990). Soil erosion and loss of organic matter further add to a decline in potential productivity each year (Ethiopia-MOA/ FAO 1984). The highland plateau, consisting of basaltic vertisols (black and red clays), is particularly prone to erosion (Huffnagel 1961; Constable 1985). Hurni (1988), for example, has shown that the areas exhibiting the highest annual rates of soil degradation are closely correlated with the worst of the two famine-prone zones outlined above—namely, the northern highland arc that covers large parts of Eritrea, Tigray, and Wollo. The high degree of degradation in these highlands plays an important role in increasing the susceptibility of farming systems to environmental crises (Dejene 1990; Stahl 1990).

It is important to note, however, that natural resource degradation and high population concentration are relatively less significant in the extreme lowlands—the second broad zone of famine vulnerability, which is largely populated by agro-pastoralists and pastoralists. More than 50 percent of Ethiopia's total land area (620 000 square kilometres) consists of "pasture" of varying quality (Caldwell 1992). In these areas, the availability of forage to support livestock, fluctuating terms of trade between livestock and grain, and mutually recognized spheres of influence based on clan groups are the key elements of the lowland economy that dictate population location and concentration (see Chapter five). Ethiopia's population density is lowest in the pasture lands of Eritrea, Hararghe, Bale, and Sidamo, where pastoral economies are strongest (Ethiopia, CSA 1991). Problems of subsistence therefore rest less on competition for land or its degradation than on the poverty of soils and vegetation and the local characteristics of rainfall.

The poor rainfall record of the drought-prone regions of Ethiopia, which are largely coincident with the two famine-prone zones, is shown in the time series in Table 2.2. These data show that low levels of mean annual rainfall in the nine provinces receiving less rain than the national long-term average are compounded by high levels of interannual fluctuation. For example, the coefficients of variation (the degree of variability around average levels) shown in column 4 of the table are highest for the provinces of Eritrea (49), Tigray (29), and Hararghe (27). By contrast, regions with the highest levels of mean

Table 2.2. Average levels and variability of rainfall in Ethiopia, 1961 to 1987 (source: derived from Webb, von Braun, and Yohannes 1992)

Provinces	Long-term average (mm)	Percentage of Ethiopia average	Coefficient of variation	Rainfall in worst year Year	Percentage of average
Arssi	872	96	16	1980	69
Bale	766	84	26	1965	69
Eritrea	398	44	49	1966	43
Gamo Gofa	747	82	21	1963	48
Gojjam	1170	128	10	1983	82
Gondar	986	108	19	1966	78
Hararghe	497	54	27	1984	49
Illubabor	1304	143	13	1965	67
Kefa	1322	145	11	1980	81
Shewa	830	91	11	1965	77
Sidamo	837	92	24	1980	51
Tigray	571	62	29	1984	44
Wollega	1210	132	20	1970	48
Wollo	837	92	18	1984	47
Ethiopia	913	100	7	1984	78

rainfall, such as Keffa and Gojjam, also display the least amount of variability, with coefficients of variation not exceeding 11. This does not mean that high rainfall areas avoid localized crises. Table 2.2 indicates that droughts have occurred in Shewa, Gondar, and Gojjam during the past 20 years. However, such droughts have been limited in scale and intensity compared with other regions.

A further element contributing to vulnerability in the drought-prone regions is their greater dependency on the short rains for their total annual production (Degefu 1987, 1988). The short rains fall between February and April, while the main rains fall from June to September. Many parts of the country benefit from short rains, which provide food (5 to 10 percent of national food production), raise grass (allowing livestock to recover weight after the dry season), and help weed control during the main rains by encouraging early growth, which can be ploughed back into the ground.

However, the regional importance of the short rains varies considerably. The central and northern highlands receive between 20 and 30 percent of their total annual rainfall from the short rains, whereas in the eastern and southern lowlands this percentage rises to 40 or 50 percent. What is more, the correlation between levels of short and long rains in the same year is low— good short rains can be followed by drought in the main rainy season (IUCN 1989).

Similarly, low rainfall during the main rains in one part of the country can be coincident with good rains elsewhere; that is, different provinces do not always experience their "worst" years of drought simultaneously. For example, column 5 of Table 2.2 shows that six of the country's provinces had their severest drought during the 1960s, only one (Wollega) was worst-affected during the 1970s, while four had their hardest year during the 1983–1986 drought. In other words, although precipitation rates in these worst years ranged from 18 percent below long-term average to as much as 57 percent below, some parts of the country escaped drought and were able to produce a harvest as normal. This differential impact of drought in space and time has important implications for the role of interregional trade as a mechanism for smoothing food supply disparities (a theme to be discussed in Chapter 3).

The lack of correlation between regional rainfall patterns illustrated in Table 2.2 supports the argument that there is no long-term pattern for Ethiopian drought episodes, be it trend, periodicity, or cycle (Lamb 1977; Degefu 1988). What is more, there is no perfect correlation between drought years and subsequent food shortages. A single year of poor rainfall in Wollo or Tigray has not invariably resulted in famine, either in the same year or the following year. Thus, although the relationship between drought and hunger is a close one (the years of 1973 and 1984 are obvious examples), it is neither a constant nor a necessary one. The link between the two is more complex, and more protracted in time, than simple associations allow.

THE LONG VIEW OF FAMINE

One of the crucial characteristics of famine is its creeping, insidious nature. The 1973–1975 famine, for example, had its origins in the late 1960s and many distress indicators were apparent long before the crisis was finally acknowledged. In 1971, Awsa District in Wollo petitioned the Governor for food aid (Goyder and Goyder 1988); a crop assessment report in 1972 indicated suffering in many parts of Wollo and warned of impending calamity on a wider scale (Goyder and Goyder 1988); and by the end of 1972, nine regional governors and administrators were requesting 46 000 metric tons of grain and 121 000 cartons of supplementary food from the Government (RRC 1985a).

By then, however, the crisis had deepened. In December 1972 the Ethiopian Red Cross was helping roughly 1000 refugees (mainly from Wollo) who had travelled south to the capital (Gill 1986). By the end of 1973, 60 000 refugees were crowded into camps in Wollo designed to cope with 20 000 (Holt, Seaman, and Rivers 1975). A survey in Wollo found that 85 percent of the population in several districts was "subsisting on less than 1500 calories per day," with 10 percent critically malnourished (RRC 1974a).

More migrants were flooding into provincial towns. A letter to the World Council of Churches from the Ethiopian Orthodox Church declared: "Wollo drought-stricken. Many dying. Mass exodus" (cited in Taube 1976). It is estimated that at least 284 000 people sought help at regional centres (Rivers et al. 1976; Hutchinson 1991).

However, it was not only Wollo that suffered. Although not centre-stage during this crisis, western Shewa was reporting 23 percent of children in 55 sampled villages below 80 percent of standard weight-for-height (ENI 1974).[3] Towns in Hararghe to the east also reported between 17 and 25 percent of children below 80 percent of weight-for-height (RRC 1974b). In other words, the crisis spread from one region to the next and from a small number of affected people to a major national crisis. Even Sidamo and Gamo Gofa in the far south were affected (Gebre-Medhin and Vahlquist 1977). In November 1973, 1.5 million people in five provinces were affected by the drought; by July 1975, there were 2.6 million people involved, spread across 11 of the country's 14 provinces (RRC 1985a).

The 1973–1975 famine may have claimed the lives of over 250 000 people, as many as 100 000 in Wollo and Tigray alone (Shepherd 1975; Gebre-Medhin and Vahlquist 1977; Fraser 1988). A survey of famine impact in Wollo found that 20 percent of the population had died in seven districts (RRC 1974a). According to Bondestam (1974), the hardest hit were Afar pastoralists, who lost 25 to 30 percent of their population.

Unfortunately, more famines unfolded during the 1980s. As early as 1977, poor rains in Tigray along with widespread crop loss and livestock mortality were noted by NGOs (Wright 1983; OXFAM 1984). Up to 2 million people were already thought to be vulnerable to famine in 1977 (Gebre-Medhin and Vahlquist 1977). As droughts and civil unrest continued into the early 1980s a government committee was established in September of 1982 to survey conditions across the country. Its report, released the following month, indicated that 3.5 million people in Tigray, Wollo, Eritrea, and Gonder would require immediate assistance to avert starvation (IDI 1983).

There were several responses. A Food Security Reserve was set up in October 1982 with a donation of 12 000 tons of grain from the World Food Programme (Clark University 1987). In December 1982, Save the Children Fund (UK) opened a feeding camp at Korem in Wollo. By April 1983 it was feeding 35 000 people, including over 900 children registered at less than 80 percent of standard weight-for-height (IDI 1983; Gill 1986). There were in addition to this camp at least 20 other feeding stations open to serve remote parts of Wollo, Tigray, and Gonder (IDI 1983). The town of Mekelle in Tigray, for example, was coping with 15 000 refugee families (Fraser 1988).

Thus, when severe drought occurred again in 1984 conditions were already ripe for a rapid transmission of food production failure to consumption failure on a massive scale. When the 1984 drought intensified across the country, the famine was in fact "already under way" (de Waal 1991). It was only in 1984 that the sequence of droughts from 1977 to 1980 began having its true effect. It was this long-term effect that BBC television cameras picked up for their harrowing October 1984 broadcast.

Figure 2.5 gives an indication of how the crisis again intensified in the worst-hit regions, particularly Wollo, and then spread outward. In Wollo, poor nutritional status, coupled with disease, resulted in considerable human mortality. In the worst months of 1985, relief agencies recorded up to 72 percent of sampled children in feeding camps at less than 65 percent of standard weight-for-height (Jareg 1987). At an emergency feeding camp in Bati (northern Wollo), the peak period of mortality was at the end of 1984 (Rahmato 1987).

It is estimated that of the deaths that occurred in northern Wollo between October 22 and December 9 of 1984, over 70 percent were those of children under 15 years of age (Demissie 1986). Despite an isolated increase in mortality (as a proportion of total camp inmates) in July 1985, which coincided with the onset of the rains, the death toll at Bati did not again reach the heights of May 1984.

Wollo shared this experience with other parts of the country. For example, in Dinki (located in the Rift Valley between Shewa and Hararghe), two out of three households experienced mortality. Other areas had relatively less human mortality during 1985, but faced destitution after the loss of productive assets. In Wolayta, oxen normally worth 300 birr were sold for 10 birr in 1985. And, in the pastoral Sidamo, many herders ate the hides of fallen livestock to stay alive.

Millions

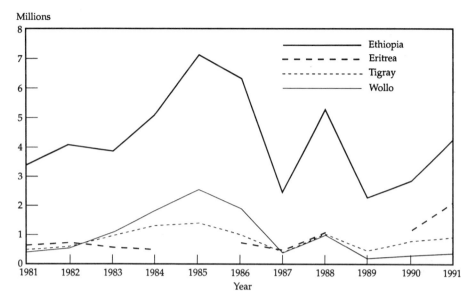

Figure 2.5. People in need of assistance in Ethiopia, by province, 1981 to 1991 (source: Relief and Rehabilitation Commission (various early warning reports); Ethiopia 1992)

Almost 8 million Ethiopians faced food shortage during the famine, and a death toll in excess of 1 million is widely quoted (Clark University 1987; Jansson, Harris, and Penrose 1987; Rahmato 1988). The latter figure (recently challenged by some as an exaggeration based on scant empirical evidence) cannot be verified (de Waal 1991). However, the magnitude of the suffering involved has been well documented, and in addition to hundreds of thousands of famine deaths, millions of people were made destitute. Such large-scale erosion of the asset base has left the poor ever more vulnerable to future crises.

The effects of the 1984/85 crisis are still being felt. Although 34 000 tons of seeds were distributed to farmers who had consumed their planting seed, the early rains of 1987 were poor, locusts destroyed much of the crop, and there was a two-month dry spell in the middle of the main rainy season. These conditions signalled further hardship. However, an appeal in 1988 for 1.05 million tons of food aid to relieve over 5 million people was immediately met by pledges, and by 1989 more than 1.1 million tons of food had been distributed, thereby reducing some of the suffering.

But the underlying conditions of vulnerability remained. After relatively good harvests in 1991, estimates of food needs for 1992 still exceeded 1 million tons (WFP 1991a; FEWS 1992). In 1993, after another good harvest, the number of people "likely to need assistance" had risen to almost 5 million, of which 1.6 million were vulnerable to shortages in the short-term (RRC 1992). Most of the vulnerable were located in Tigray, Hararghe, Wollo, Borena, and Wollega (RRC 1992; FEWS 1993a). The United Nations therefore launched another international appeal for donations, this time for US$299.9 million to protect and rehabilitate up to 5 million people "suffering from the effects of civil war and drought" (Barnhart 1993).

The continuation of famine vulnerability into the mid-1990s involving millions of people has been widely ascribed to the climax of the civil war in the northern regions and to pockets of drought that continue to hamper growth in many other parts of the country. But, are these the only, or even the principal, explanatory factors to be considered? We have seen in this chapter that many elements combine over an extended period of time to result in an "event" that captures a government's (and the world's) attention. The next chapter therefore explores the complex issue of famine causation, with a focus on major contributing factors in both the short and the long term.

NOTES

1. Eritrea's independence from Ethiopia was formally recognized by the Ethiopian authorities and the United Nations in May 1993. However, because most of the analysis presented here relates to the pre-1993 period, Eritrea is generally treated in the text and in maps as an administrative province of Ethiopia.
2. It should be remembered that the lifting (in the 1990s) of policy restrictions on personal travel and migration between provinces will allow for much greater horizontal, and also vertical, movement of population than in the previous two decades. This may have an impact on the nature and location of pre-1990 population concentrations.
3. Weight-for-height is a commonly used measure of wasting (short-term malnutrition). It compares a child's weight and height against an international standard (provided by NCHS 1977). If a child is below 80 percent of the international standard weight for a given height then the child is considered to be seriously malnourished.

3 The Labyrinth of Famine Causality

Despite standard images of suffering associated with famine, there is no standard explanation for its occurrence. Multiple factors are at work to generate conditions conducive to societal breakdown and individual starvation. Multiple actions are required to prevent it.

This intrinsic ambiguity has provided fertile ground for debate. During the 1960s, famines in newly independent African states were frequently discussed in the light of theories about colonial exploitation and the dependency of "periphery" countries on more developed western economies (Bhatia 1967; Amin 1976; Franke and Chasin 1980). Widespread drought in the early 1970s generated an outpouring of literature on the "people versus nature" theme (Bryson and Murray 1977; Schneider 1977). The 1980s focused on issues relating to the breakdown of markets and to the collapse of household entitlements (Sen 1981; Vaughn 1987; Curtis, Hubbard, and Shepherd 1988). The 1990s appear to be taking on all of the above, with attention being paid to the interaction between multiple elements that are no longer seen as alternatives but as complements (Shipton 1990; Currey 1992; Watts and Bohle 1992).

There exist already numerous taxonomies that categorize the factors that play a part in famine causation. Most taxonomies differentiate between factors according to a time dimension. For example, Jelliffe and Jelliffe (1971) refer to initial, aggravating and ameliorating factors spanning different periods from decades to months. Others use the terms baseline and current indices (Downing 1990), proximate and intermediary variables (Swift 1989), and underlying processes, immediate causes and direct consequences (Kates and Millman 1990).

What each of these phrases means is that some causal factors operate as processes over a long period of time while others are more discrete short-term events. Population growth and ecological degradation, for example, are widely seen as processes of a Malthusian nature that increase the likelihood of food supply failure in the event of a drought or economic crisis (Constable 1984; Hurni 1988; Dejene 1990). These are initial, underlying or baseline factors. By contrast, policy mismanagement, military conflict, drought, and market collapse are factors of a more current or proximate nature that can trigger a crisis where only the threat of crisis existed before (Degefu 1988; Winer 1989; Lancaster 1990).

This chapter focuses on the policy, conflict, drought, and market failure issues, four more immediate factors that are most commonly raised in explanation of famines in Ethiopia during the 1970s and 1980s. We do not seek to pin blame on one factor more than another. Rather, the aim is to accept at the outset that all four elements are important and therefore to examine them and discuss their interaction. We begin by documenting the problem of declining food availability in Ethiopia, and then address the four causal factors one at a time.

TRENDS IN FOOD PRODUCTION AND AVAILABILITY

Cereal production and availability have been declining in Ethiopia since the 1960s. Figure 3.1 shows that cereal production per person has dropped by an average of 4 kilogrammes per year. The decline has not been smooth and uninterrupted over the entire period. Certain years, such as 1975 and 1987, saw large increases in production that may have been associated with above-average rainfall, policy changes, and even with changes in the way production estimates are made.[1]

What is more, the decline has not been smooth across all regions. Because of Ethiopia's diverse agro-climatic conditions there are variations in the relative importance of cereals and root crops by region, and a strong geographical

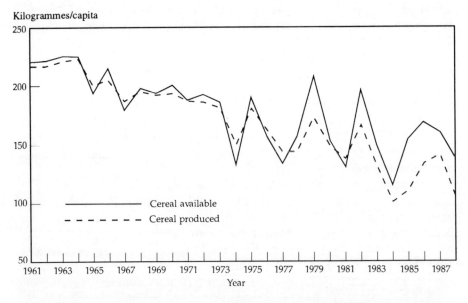

Figure 3.1. Cereal production and availability per capita, 1961–1988 (sources: derived from Webb, von Braun, and Yohannes 1992; Ethiopia-CSA 1987a,c; 1988a,b; 1989a,b)

concentration of cereal production. For example, the regions growing most teff (*Eragrostis tef*—an indigenous species of short-stalked grain) are Shewa and Gojjam. These two provinces provide almost two-thirds of the country's supply of teff. Maize, on the other hand, is produced in large quantities in Keffa, Sidamo, and Wollega as well as in Shewa and Gojjam.

However, if one takes total cereal output, the concentration of production is unequivocal. Shewa, Gojjam, and Arssi were together responsible for the bulk of cereal production through most of the 1980s and early 1990s, both in terms of share of national production (typically greater than 50 percent), and in terms of production per person by province. The areas producing the least grain are Gamo Gofa (which depends largely on root crops), Bale (which is sparsely populated), and Illubabor (a major coffee-producing province).

Despite the existence of these pockets of higher than average production, the overall trend of cereal production has been down. Since food availability in Ethiopia is strongly determined by the country's own production of cereals (having little capacity to purchase food on international markets), cereal availability has been declining at an average of 3.3 kilogrammes per person per year.[2] In the late 1980s, the country was producing less than 150 kilogrammes of cereal per person. The level required for a minimum subsistence diet is approximately 240 kilogrammes per person per year (FAO 1990). For the sake of comparison, only 20 percent of Niger's land area receives enough rain for unirrigated farming to take place, yet it produced an average of 330 kilogrammes per person per year between 1960 and 1990—at least double Ethiopia's output in the early 1990s (Webb 1992).

Of course, cereals are not all that matters. The cultivation of other kinds of food, such as pulses (varieties of beans) and root crops (such as cassava), serves to supplement cereal production (Vosti 1992). Unfortunately, these crops have fared little better than their cereal counterparts. Although few data exist on root crop production, pulse and oilseed production in 1989 stood at less than 75 percent of its 1979 level (Gutu, Lambert and Maxwell 1990). As a result, the availability of cereals plus pulses and oilseeds still declined by an average of 2.7 kilogrammes per person per year up to the early 1990s.

The fact that since 1977 there has generally been more cereals available than were domestically produced is explained by imports, both commercial and food aid. Given chronic foreign exchange constraints, commercial imports only reached 215 000 tons during 1985. Food aid, on the other hand, increased significantly after 1984 and made an important contribution to food availability in the critical years of 1985, 1988, and 1991. In each of those years aid deliveries reached roughly 1 million tons of cereal, representing 15 to 20 percent of national production (Aylieff 1993).

However, it should be remembered that with a total population of over 45 million people, food aid contributed during crisis years cannot meet all need. Assuming that all of the 1.2 million tons delivered in 1985 reached the mouths of the 6.9 million most vulnerable people, then aid represented 175 kilo-grammes of cereal per starving person. This would have been sufficient to keep each vulnerable person alive (albeit at a suboptimal level of food consumption) for almost 1 year.

But two things are certain; first, most vulnerable people were living in areas that suffered higher than average shortfalls in food production during crisis years, as well as in the long-term average decline; and second, not all the food aid reached the most vulnerable people. The following sections examine reasons for both occurrences.

MILITARY CONFLICT

After famine, Ethiopia is perhaps best known for the longest running civil war in recent African history—a conflict often blamed for keeping food out of the mouths of the hungry. The independence struggle pursued by the Eritrean People's Liberation Front against successive Ethiopian governments came to an end in May 1991 after almost 30 years. The result was admittance to the United Nations as a recognized state in May 1993.

But, there has not been only one war. To the protracted conflict over Eritrea must be added the northern based resistance to central Ethiopian authority waged for some 15 years by the Tigray Peoples' Liberation Front (Keller 1988). This movement of opposition to the totalitarian autocracy of President Mengistu spread in the 1980s to include other rebel factions in Gonder and Wollo (Tiruneh 1993). This loose coalition of rebel forces marched into Addis Ababa in 1991 to oust the Mengistu Government and assumed control of the country.

There was also an international war in 1977 when Somalia invaded the Ogaden and was repelled only on the outskirts of Dire Dawa. But, this military success and moral boost strengthened the central government for a few years only. In the 1980s, central power became increasingly eroded as the northern conflicts spread south and westward. This strengthened the lure of ethnic "nationalism", giving rise to rebel action by Oromo groups in Hararghe and Sidamo (Tiruneh 1993). Furthermore, as government control over many rural areas diminished, bandit groups became increasingly active in Shewa and the Ogaden. In other words, localized fighting in the north fuelled insecurity elsewhere in the country, which generated fighting across large swathes of the country—a typical sequence of events in the break-up of empires (Gibbon 1787; Churchill 1956).

Most of the conflicts since the 1960s had their roots in the process of Empire building that produced the Ethiopian State only a century ago. Menelik, King of Shewa, extended his dominion over northern, central, and southern regions in the 1880s and became Emperor in 1889. His strength restricted the Italian army to colonization of Eritrea in the early 1890s when they were seeking to control the whole of Abyssinia. Menelik died in 1913 and his successor, Ras Tafari (succeeding as Regent but later to become Emperor Haile Selassie), faced a major uprising in Tigray and Wollo in 1916/17 as disaffection with centralized rule from Shewa grew (Thesinger 1987; Holcomb and Ibssa 1990). This uprising was ruthlessly put down, but represented a precursor to rebel activity six decades later (Tareke 1991).

Federated to Ethiopia by the United Nations General Assembly in 1952, Eritrea was formally annexed by Ethiopia in 1962, and thereby lost the nominal autonomy that it had gained since the expulsion of the Italians during the Second World War (Cliffe and Davidson 1988). Armed resistance to this annexation began immediately, and the scale and intensity of fighting grew steadily, and well beyond Eritrea's borders, until 1991.

The costs of conflict have been enormous. In economic terms it has been argued that the internal wars were significantly more costly than the external war with Somalia (Henze 1984). Despite reports that the Soviet Union (a new ally after the war with Somalia) had already armed Ethiopia with military equipment worth up to US$3 billion, President Mengistu allocated 46 percent of the Gross National Product for 1984 towards building the largest standing army in Sub-Saharan Africa (Bush 1985; Lemma 1985). In 1988, President Mengistu reported that the Government was spending over US$700 million per year on the military (Buchanan 1990; Horn of Africa Report 1990; Lancaster 1990). As a result, the defense budget was officially increased from 11 percent in 1974/75 to 37 percent in 1990/91 (Kloos 1991). The latter figure has been dismissed as a gross underestimate (Chole 1989).

Such huge war expenditure inevitably drained the economic coffers, diverting investment away from development activities. The health budget, for example, fell from 6 percent in 1973/74 to only 3 percent in 1990/91 (Kloos 1991). Most other sectors of the economy were similarly bled of resources. Diversion of resources also occurred in the food aid sector. There is little doubt that large quantities of food were used to feed the army during the mid-1980s famine (Lemma 1985; Kaplan 1988; Keller 1992). This made the task of saving lives and of subsequent rehabilitation that much more difficult.

Donors were faced with the conflicting desires of wanting to cooperate with the Government in order to reach famine-affected populations in Government-held territory, and not allowing the regime to use food as a weapon of war. This was partly overcome by organizing "corridors of tranquility" through which donor-controlled food convoys could reach famine areas in relative safety, and by bringing large food shipments to contested regions through territories already controlled by rebel movements (Hendrie 1989).

Food, however, was not the only assistance that famine-prone regions required. The compression of government investment in nonmilitary sectors meant that the country's economic and physical infrastructure was not just damaged by the war, its reconstruction and expansion were seriously compromised for almost two decades. This included, for example, the roads and medical facilities required to deal with and evacuate many thousands of casualties.

Although difficult to calculate, and always open to dispute, estimates of war casualties and physical damage in Ethiopia are typically very high. Nationwide conscription drives (dating from 1976) brought Government forces up to a level of almost half a million men by the end of the 1980s. Between 1976 and 1982 such drives were carried out on an irregular basis, but the decision in 1981 to formalize national military service for the "entire working people" resulted in annual conscription campaigns across the country (de Waal 1991). It should be underlined that in 1984 the total adult male population of Ethiopia aged between 18 and 40 was 5.8 million (E. Richardson, personal communication 1993). Thus, the army at that time contained almost one in ten of the male working-age population.

The numbers of conscripts, opposition forces, and civilians killed during the conflict cannot be known, although estimates abound. For example, Wolde-Giorgis (1989) attributes a figure of 250000 deaths to the war in Eritrea from 1967 to 1988. Cliffe (1989) is more circumspect in proposing a total of 5000 dead and imprisoned (on both sides) between 1986 and 1989. Davidson (1993) suggests a compromise at 65000 dead and 100000 homeless.

For Ethiopia as a whole, Sivard (1991) puts the total number of war dead between 1974 and 1990 at 609000, more than 500000 of them civilians. Official estimates are higher. On the basis of its own data, the Transitional Government of Ethiopia has proposed that 150000 members of rebel movements and 500000 civilians died during Mengistu's rule (Eliassen and Eriksson 1991).

These figures are for the dead. The wounded and handicapped live on. The 1984 census already reported the existence of 40000 amputees in the country (Ethiopia-OPHCC 1991). On top of that, roughly one-third of the 300000 demobilized soldiers returning home in late 1991 were thought to be injured or disabled (CRDA 1991).

These high human casualties have an important bearing on food production in the country. It has been estimated that the conflict within Eritrea alone cost between 65000 and 95000 tons of lost food production per year (Bondestam, Cliffe, and White 1988). Therefore, good rainfall coupled with peace could raise Eritrea's cereal production by an estimated 50 percent (CDS 1992). The direct impact of conflict is therefore great. Cliffe (1989) calculates (again just for Eritrea) that between 1986 and 1989, 23000 hectares of land were rendered unfit for cultivation, 2500 homes were destroyed, and 44000 animals were lost to their owners. Similar impacts of conflict were felt in many other parts of Ethiopia. Food production was compromised in highland areas by coercive conscription and fear of looting (Kloos 1991). Output in Tigray and

Wollo was affected by widespread losses of plough oxen as well as by the mining of fields.

In regions not affected by conflict, a lesser but still important effect of the war was the levying of special taxes on smallholders to support the war effort. Most parts of the country never experienced any bombing, entrenchment, or battles. However, in addition to an annual agricultural tax of 20 birr (roughly US$10), a membership fee of 5 to 10 birr for obligatory membership of a Peasant Association, and other association and social service support fees (often exceeding a total of 15 birr), a "voluntary" contribution was requested as of 1988 for support of the "territorial integrity of the Motherland."[3] Thus, in 1988/89, many households were expected to pay an average of around 50 birr per year in contributions to the State.

War, the most brutal of the horsemen of the Apocalypse, has ridden rough-shod over Ethiopia for decades. The damage is both obvious and incalculable. Yet, although war bears a large share of the blame for food insecurity and famine, it does not explain all of the problem. Famine is not necessarily the riding partner of War. It should be underlined that up to 50 percent of the people identified as vulnerable to famine during the late 1980s and early 1990s were located outside of the main zones of conflict (see Figure 2.5 in the previous chapter). In these areas, chronically low and variable farm produc-tivity, a lack of valuable household assets, low income levels, and widespread shortfalls in food consumption certainly reflect a lack of public investment in rural development (due largely to military expenditure), but other factors are also important.

DROUGHT AND CROP FAILURE

Drought represents a seasonal moisture deficit significantly below long-term average levels for a given locality. In the United Kingdom, drought is usually associated with insufficient rainfall to replenish reservoirs. A period of 15 or more consecutive days (at any time of the year), each receiving less than 0.01 inches of rain, is typically defined as a drought in England (Barry 1969). If drought extends beyond 30 days water rationing may come into effect, with the use of hose-pipes for washing cars or sprinklers for watering lawns being prohibited for a short period of time.

In Africa, drought is linked to crop failure during the principal growing season. In parts of Ethiopia, drought is defined by farmers as the absence of rain when required for seed germination, plant fertilization, and crop growth (Wolde-Mariam 1991).[4] The lack of reservoirs for water storage and of irriga-tion technology makes the protection of crops during drought practically impossible.

Twenty-one countries in Sub-Saharan Africa experienced a severe drought in 1984/85. In each case, crop production was seriously compromised.

However, only a handful of those countries suffered famine (Harrison 1988). Similarly, most parts of Ethiopia experienced drought in 1984/85, but not all regions were affected to the same degree, nor did they all suffer famine. The drought–famine link is therefore not a perfect one.

It was shown earlier that rainfall in one region rarely coincides exactly with rainfall in other regions. Figure 3.2 shows rainfall for Keffa (a region of high agricultural potential) with that of Hararghe and Wollo (two drought-prone regions). The figure shows that rainfall was exceptionally low in all three provinces in bad years such as 1965, 1973, and 1980. However, in 1984, a catastrophic year for Wollo and Hararghe, Gojjam received the best rains in a decade, well above the national average. The Ethiopia total average line shows troughs for the years when the worst droughts were experienced. But the troughs are fairly shallow because figures for the drought regions are cancelled out by those from regions such as Keffa.

The figure also shows that interannual fluctuations are larger, and drought sequences are longer, in drought-prone regions than is typical for the country as a whole. This confirms that an isolated drought is rarely a dangerous drought. For Wollo and Hararghe, the 1973/74 drought was merely the last

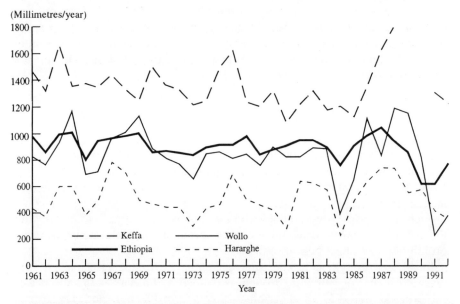

Figure 3.2. Mean annual rainfall for Keffa, Hararghe, and Wollo provinces, and for Ethiopia (total), 1961 to 1992. Note that the geographic coverage of rainfall measurement sites declined sharply from almost 500 stations in 1987, to 245 in 1989, to only 53 in 1992. Thus, the data from 1989 onwards should be treated with caution (source: based on data obtained from the National Meteorological Services Authority, Government of Ethiopia)

and worst of a sequence of five to seven years of declining rainfall. Similar drought sequences preceded the crisis years of 1979/80 and 1984/85. In other words, it is only when one poor year follows others that droughts take on unmanageable proportions (Hay 1986; Corbett 1988; de Waal 1988). The disasters of 1973/74 and 1984/85 did not happen suddenly; they were the culmination of a process spread over many years.

Drought therefore affects crop production in two principal ways. The first is that it may affect cropping patterns over the longer term through farmers' perception of the likelihood of drought sequences in their region. That is, they may choose to cultivate low yielding more drought-resistant crops, such as millet, rather than higher yielding but drought-sensitive crops, such as barley. The second effect is through a direct impact on crop yields.

In Ethiopia's case, the first effect appears minimal, at least at a national level. Droughts of varying severity and duration have been common in Ethiopia for centuries. Farmers acknowledge the risk of drought once in every four to six years by diversifying their crops and their sources of income. However, the risk of drought does not seem to cause farmers to significantly adjust the types of crop grown across years or regions. Such patterns are determined more by local agro-ecological conditions and by relative price incentives.

For example, there were no big changes in the proportions of different cereal crops planted at a national level between 1970 and 1989. In 1989, teff was the dominant cereal, occupying 30 percent of Ethiopia's total planted area (Gutu, Lambert, and Maxwell 1990). The next most important cereals were maize and barley, each accounting for approximately 20 percent of total area. Millet and oats were the least important cereals, together accounting for less than 5 percent of area. Wheat and rice took up the remaining land.

Most of these shares remained largely unchanged over a 20-year period. The one exception was a reduction between 1970 and 1989 in the area share of sorghum (from 18 to 13 percent), paralleled by an upward trend in maize (from 16 to 20 percent). The increased focus on maize was due to three main factors: first, the loss to disease of root crops, such as enset (*Ensete ventricosum*, an indigenous tuber) in southern regions; second, to the distribution of shorter-maturing varieties by the Institute of Agricultural Research; and third, to the price ratio between maize and sorghum (consistently in favour of maize during the 1980s), which encouraged cash-poor farmers to replenish lost seed stocks with the cheaper cultivar.

The direct impact of drought on crop yields is stronger. The relationship between rainfall quantity, distribution, timing, and resultant crop yields is complex. Soil type, soil depth, location of plots, and soil temperature all have an effect on yields and total production (Henricksen and Durkin 1985; Wolde-Mariam, personal communication, 1992). Nevertheless, one expects a simple link between total level of rainfall and crop conditions. Although many factors other than rain determine output (such as economic incentives, financing, and input availability), a sharp decline in rainfall plays an overriding role in cereal production where farming is carried out at a low level of technology and inputs—the case across most of Africa.

The simple link between rain and yields can be calculated using a statistical regression model that matches cereal yields and production with total rainfall.[5] The results presented in Table 3.1 for the years 1961/62 to 1988/89 show that one can indeed explain a relatively large share of the actual variance in both cereal yields and production over time with regard to levels of rainfall. The explanatory power of the R^2 terms exceeds 75 percent, underlining the fact that in the absence of yield-improving technologies and inputs Ethiopian agriculture remains extremely dependent on rainfall for its success or failure.

These results, and others focused on the response of individual cereal types to rainfall variability, suggest that a 10 percent decline in rainfall below the long-term national average results in a fall in all cereal yields of an average of 4.2 percent (Webb, von Braun, and Yohannes 1992). Of course, the effects on yields and production vary by region. Just as droughts are not correlated across regions, production in one province is not correlated with production of any other province. There is, therefore, variety within the country in terms of production variability.

It can therefore be concluded that relationships between domestic production and food availability are strong in Ethiopia at both national and provincial levels. Drought has to be seen as an important factor contributing to food supply shortages in Ethiopia.

But, as already argued, neither drought nor food supply in themselves determine whether a famine will occur. The importance of drought–

Table 3.1. Rainfall–cereal production relationships in Ethiopia: regression analyses for 1961/62 to 1988/89 (t values in parentheses; number of cases = 28)

Dependent variables	Explanatory variables				R^2	F value
	$RAIN_t$	$RAINSQ_t$	$DUMMY_t$	CONSTANT		
$CEPROD_t$	25075.4	−12.259	−1448382.9	−6978123.5	0.77	26.4
	(2.0)	(−1.8)	(−8.3)	(−1.2)		
$CEYILD_t$	6.587	−0.0033	−340.2	−2079.8	0.76	24.7
	(2.2)	(−2.1)	(−8.1)	(−1.5)		

$CEPROD_t$ = total cereal production in metric tons in year t.
$CEYILD_t$ = cereal yield in kilogrammes per hectare.
$RAIN_t$ = annual rainfall (country-weighted averages of regional rainfall using production shares as weights).
$RAINSQ_t$ = RAIN squared.
$DUMMY_t$ = dummy variable for separation of production series before and after new statistics system (1978).

production and production–availability linkages to famine is mediated by domestic policies and by the functioning of domestic markets. The next section therefore examines the broad outlines of agricultural policy in Ethiopia during the 1970s and 1980s, before moving on to consider the question of markets and price behaviour.

AGRICULTURAL POLICY OF THE MARXIST ERA

The agricultural sector is crucial to Ethiopia's economy. It employs over 85 percent of the active population and accounts for almost 50 percent of GDP. Agricultural products constitute 85 percent of total exports, deriving 90 percent of total export earnings (Belete, Dillon, and Anderson 1991). Coffee, the country's largest export, accounted for more than 50 percent of foreign exchange earnings during the 1980s, with exports exceeding 94 000 tons in 1988/89 (National Bank of Ethiopia 1989).

Despite the importance of this sector, agriculture has for more than two decades shown disturbing signs of decline. Real Gross Domestic Product, driven by agricultural output, fell from 4 percent per year in the 1965 to 1973 period to an average of 1.5 percent per year between 1974 and 1991 (USAID 1993). Coffee revenues, already compromised by the collapse of the International Coffee Agreement in 1989, fell to only 36 000 tons in 1992 (IFAD 1989; USAID 1993). Domestic production was falling even in non-drought years.

The emperor Haile Sellassie was overthrown in 1974 largely as a consequence of famine in Wollo and Tigray, which spread unabated for several years. Coming to power in the wake of mass starvation, the Revolutionary Government committed itself early to addressing the issue of hunger through raised agricultural output. The principal articulation of this commitment was the implementation of three major policy initiatives aimed at state-led growth in agriculture: land reform, the aggregation of production units (cooperativization and association), and a narrow sectoral and geographical concentration of investment.

Despite low overall investment in the agricultural sector (averaging 9 percent of total government expenditure from 1974 to 1984), these initiatives were responsible for fundamental changes in the rural economy (Diakosavvas 1989; IFAD 1989; Pausewang et al. 1990). Although discarded during the policy reorientations of the 1990s, a brief outline of the former role of such policies, still pursued in a number of variants elsewhere in Africa, is pertinent to an understanding of trends in agriculture during the 1980s.[6]

Land Reform

In an attempt to redistribute income and stimulate agriculture, private ownership of land was abolished in 1975, a move that changed the face of farming.

Although tenure arrangements, cropping patterns, and technology varied across the country, agriculture before 1975 was dominated by the needs of a landed gentry, the church establishment, and a quasi-feudal aristocracy.[7] The reform measures transformed this structure and laid the foundations of a three-tier system.

The largest component of this system was (and remains) the smallholder sector. Individual households, which are responsible for about 90 percent of national production, were granted access rights to a maximum of 10 hectares for private production (Brüne 1990). In practice, household holdings averaged 1 to 2 hectares. Over 5 million households were organized into some 20 000 Peasant Associations. These associations controlled the allocation and use of land, each being responsible for as much as 800 hectares.

Almost 40 percent of the Peasant Associations were organized into new villages by the end of the 1980s (IFAD 1989). That is, the traditional pattern of dispersed habitation was rejected by the government in favour of collective units. Families were expected to leave their old homes and to build new ones on predesignated plots.[8]

Land reform went some way towards equalizing access to land and providing security for the landless. Such measures were welcomed most (and therefore implemented more smoothly) in the central and southern regions, formerly characterized by extensive absentee ownership of land, tribute farming, and large-scale commercial agriculture (Cohen and Weintraub 1975; Abate and Teklu 1979; Blackhurst 1980). For example, some landlords in Arssi province had controlled over 1000 hectares before 1975, which they farmed through tenants who paid up to two-thirds of their cereal harvest in rent. Yet, land reform was just the first step in the new political agenda. Other elements were needed to support the evolution of the proposed collectivist agriculture.

Aggregation of Production

To create the second tier of the new rural system, Peasant Associations were encouraged to work toward the formation of cooperatives. The first step was the organization of Service Cooperatives. These organizations were designed to sell farm inputs, provide storage and processing facilities, offer low-interest loans, and facilitate the sale of local produce (Cohen and Isaksson 1987a). A total of 3600 cooperatives were serving 4.4 million households by the end of 1988 (IFAD 1989).

The next stage was the formation of Producer Cooperatives. These cooperatives pooled land, labour, and other resources in an attempt to capture economies of scale. In order to attract members, the cooperatives offered low taxes, interest-free loans, and priority access to inputs and consumer goods. By 1988, almost 4000 Producer Cooperatives had been founded, comprising

302 600 households (Walters 1989). However, in that year they were responsible for only 5.5 percent of national cereal production (Cohen, Goldsmith and Mellor 1976a; Griffin and Hay 1985; Gutu, Lambert and Maxwell 1990). Empirical evidence of the contribution (if any) of such associations to hunger and poverty alleviation is lacking (Teka 1984; Pankhurst 1985; Desta 1990). Unfortunately, because most cooperatives were dissolved after the 1991 change of government the opportunity to gather such evidence in order to compare Ethiopia's experience with that of Tanzania and Zambia has all but disappeared.

The third, and smallest, component of the farming system was made up of State Farms. Of an estimated 750 000 hectares of private commercial farms functioning before 1974, 67 000 were converted into State Farms that were operated (from 1979) by a new Ministry of State Farms (Cohen and Isaksson 1987b). The remaining portion was dismantled and used either for settling the landless or for assimilation into adjoining peasant associations. By 1989, State Farms had grown to occupy a total area of 220 000 hectares (Walters 1989; Brüne 1990). However, despite large investments and running costs, these farms were responsible for only 4.2 percent of main season cereal production in 1988/89. This relative inefficiency has been widely criticized (Wörz 1989; Winer 1989; Rahmato 1990b).

Concentration of Investments

The third trend of the 1970s and 1980s was a concentration of investment in the provision of improved inputs to selected farmers in high-growth-potential regions and to the State Farms. The Mengistu Government rejected pre-1974 strategies focusing resources on capital-intensive mechanized farming (Aredo 1990; Girgre 1991). For example, from 1968 to 1972 more than 76 percent of capital expenditure in agriculture (9 percent of government spending) went to the commercial (export) sector, compared with 13 percent for the smallholder sector (IFAD 1989). By contrast, the Ten-Year Perspective Plan (1984–1994) proposed to raise agricultural expenditure to 30 percent but also allocated 37 percent of this to the smallholder sector (IFAD 1989).

However, although this was an improvement for the smallholder sector, State Farms still received the bulk of agricultural spending until 1990. Furthermore, not all small farmers benefited from the increased smallholder allocation. Although much pre-1985 investment attempted to improve basic infrastructure for input delivery nationwide, the period from 1985 up to the mid-1990s was dominated by plans to concentrate foreign exchange and skilled managerial inputs, totalling US$85 million, on a small number of farmers in high potential regions (the Peasant Agricultural Development Project [PADEP]). For example, the 1986–1989 Public Investment Programme selectively concentrated its expenditure (much of it related to PADEP) on

181 "surplus-producing" districts in the west of the country. Similarly, between 1987 and 1989, donors provided almost US$237 million for agriculture projects; one-third of the total was accounted for by PADEP and very little of the remainder funded projects in drought-prone regions (IFAD 1989).

There has been some questioning of this approach, including criticism of the justifications used for PADEP: for example, reference to yield increases achieved by Ministry of Agriculture station trials rather than farm trials, and only a superficial analysis of the experience of the 1960s Chilalo Agricultural Development Unit (Kifle 1972; Koehn 1979; Gamaledinn 1987; IFAD 1989). Nevertheless, many donors support this initiative, and the principle of expanding output where potential is highest currently guides most project activities. This principle is not misplaced in simple economic terms. Yet, without parallel investment in addressing food insecurity among the poor in drought-prone regions, the net benefit of raising productivity in growth-potential regions may be compromised in the longer term.

Impact of Marxist Policy

Public commitment to improving the welfare of the rural poor by revitalizing agriculture served as the justification for policy experiments of the Marxist era. Most were aimed at changing the structure of productive relationships between land, labour, capital, and output. However, the effects of such policies on productivity in general, and on levels of food self-sufficiency in particular, have been limited.

Although land reform and cooperativization brought smallholders into a new relationship with the state (and with each other), thereby supplanting the role of landowner and church, these measures did little to remove long-standing constraints to improved farm productivity (Brüne 1990). Thus, a positive effect on output has not been observed. For example, the impact of land reform was small because, (1) quality of land and degree of land fragmentation were rarely considered, (2) area cultivated per capita has remained small, (3) redistribution was not supported by increased security of tenure, and (4) access to land is only one variable in the production function—access to oxen and inputs was largely unaffected by the reform.

The process of social and economic aggregation was intended to improve the economic basis of land use resulting from reform and redistribution. However, new organizations rarely met their potential for stimulating rural growth. First, they were perceived as tools for enhanced state control of the rural environment, giving preferential treatment to party members. Second, cooperatives suffered from a lack of resources (improved inputs and credit) to meet demand (Rahmato 1990b; Lycett 1992). Third, price incentives necessary to raise productivity were lacking, thereby limiting any potential gains made possible either by land reform or cooperativization.

Thus, there is a broad consensus that although the structures of rural society and production that pre-existed 1974 needed changing, the policies pursued between 1975 and 1990 did not succeed in raising national or household food security on a large scale (Magrath 1991). The smallholder sector did not respond to the altered policy environment, largely because it could not. The causes of poverty had not been removed, and with the mere stabilization of consumption levels (an important priority for a large proportion of households), few could raise a large marketable surplus.

At the same time, the State Farms, consuming most agricultural investment but returning less than 5 percent of total production, was far from reaching its goal of alleviating the country's food problems (Wörz 1989; Rahmato 1990b; Mirotchie and Taylor 1993). The pre-1975 agrarian economy was hardly dynamic (Belete, Dillon, and Anderson 1991). But after the policy changes of the 1975 to 1991 period the rural economy was stagnant, if not declining.

Policy Adjustments of the 1990s

By 1990, the shortcomings of previous policies were openly acknowledged. A number of new strategy plans were elaborated in the late 1980s that sought to clarify the role of government in famine relief and long-term prevention, and the scope for a new focus on smallholder agriculture. For example, 1989/90 saw the inception of a National Conservation Strategy and a National Disaster Prevention and Preparedness Strategy (DPPS).

The principal components of the DPPS, in particular, deserve to be highlighted here. They comprise commitments toward (1) emergency legislation designed to delegate responsibilities and speed up responses to crises; (2) institution building, that is, a strengthening of the planning and response capacities of relevant government organizations; (3) investment in enhanced information systems (which will guide the appropriate crisis response); and (4) preparation of interventions to enhance institutional readiness for action. The fifth component is a commitment to a streamlined integration of the first four initiatives with longer term development programming (ONCCP 1989).

The increasing convergence of opinion expressed in such documents on the causes of low productivity and related food insecurity was highlighted by a policy reorientation announced in March 1990. Termed a "New Economic Policy," a package of reforms was proposed that merged many of the elements suggested in the various strategy papers. Key changes included legalization of labour hiring for private production (and, hence, of a free rural labour market), removal of the control stations regulating the movement of grain across provincial borders, abolition of quotas requiring the sale of grain to the Agricultural Marketing Corporation at fixed prices, removal of restrictions on private investment, and denationalization of uncompetitive enterprises (Baker 1990; Belshaw 1990; IGADD 1990; Pausewang et al. 1990).

These reforms signalled a concern with the structural problems of low land and labour productivity, and associated poverty and famine vulnerability. However, they came too late to forestall the downfall of the Mengistu Government in 1991. This change initiated a far-reaching debate about the policy requirements of any subsequent democratically elected authorities.

Despite the expected opposition between proponents of market versus state control of the economy, agreement emerged during 1992 in three key areas: namely, the need for (1) more secure, individual, tenure rights to land and other natural resources; (2) a refocusing of agricultural investment towards the smallholder sector, and (3) a fully implemented political, as well as economic, liberalization of the rural environment leading to genuinely participatory planning and implementation of development activities (Manyazewal 1992; Richburg 1992). Interestingly, these are the same three issues that dominated debate in the 1970s on how to overcome the structural causes of low agricultural productivity inherited from the imperial era (Bondestam 1974; Koehn 1979; Cohen, Goldsmith, and Mellor 1976b; Brüne 1990).

The debate continues. This in itself is a positive development over conditions prior to 1991. What is more, prospects for growth appear strong. In 1992/93, an Emergency Recovery and Reconstruction Programme (ERRP) was financed by 11 separate donors to a sum of almost US$700 million (USAID 1993). This was accompanied by the rescheduling of Ethiopia's external debt, and a halving of the service burden on US$600 million in hard currency debt and on 800 million rubles (USAID 1993). By 1994, a US$250 million Structural Adjustment Credit from the World Bank came on line, matched by US$550 million in parallel financing by other donors.

Supported by these grants and credits, the Transitional Government of Ethiopia implemented a series of structural economic reforms agreed with the International Monetary Fund and the World Bank. These included most of the proposals laid out in the New Economic Policy outlined above, such as the decontrol of retail prices, currency devaluation, the lifting of restrictions on transport tariffs, dissolution of many public enterprises, a lowering of income tax rates and liberalization of investment and labour laws to promote private initiative (TGE 1992; World Bank 1993).

With these measures underway, agricultural output was expected to exceed 9 percent per annum in the mid-1990s, based principally on a 7 percent expansion in area cultivated (USAID 1993). However, the debate over the future takes place in the context of renewed uncertainty. Although the change of government and cessation of war have resolved many problems, it is unclear what future directions the new government will take. The key issue of land tenure was conspicuous in its absence among major policy changes of the mid-1990s. The policy of redrawing internal administrative boundaries along ethnic lines has raised fears of "ethnic cleansing" following the experience of former Yugoslavia. And, the poor remain in a state of high vulnerability to famine, which raises the danger of political instability in the context of future crises.

Can "the market" adequately protect the poor in such circumstances and where market infrastructure is chronically deficient and food prices continue to fluctuate? This issue is addressed in the next section, which examines the crucial relationship between prices and food availability.

PRICES AND MARKETS DURING FAMINE

An understanding of market and price behaviour during crises provides insights into the relative merit of "free" versus "state-guided" market operation during famine. The latter remains a crucial and still much-debated policy issue in many parts of Africa. In an environment of well-integrated markets (for food, labour, and capital), local production shortfalls result in food price increases and a decline in real wages.

A sudden collapse of purchasing power—in the context of production failure and a parallel decline in employment—may not force sharp price rises because of a fall in effective demand. If a region experiences frequent famines the potential for relying on the market to mitigate suffering becomes increasingly limited, because the asset base of most households is progressively eroded. In such a case, prices become a poor barometer of actual need because effective demand is so weak. Thus, food prices do not necessarily send appropriate signals to private traders, nor do they necessarily serve well as a guide to the timing of public intervention.

Policy Restrictions on the Market

Interregional trade in Ethiopia during famine years of the 1980s was impaired by government trade restrictions. As a result, famine in one region was able to coexist with a reasonable balance of food supply and demand in other regions.

Before 1975, government intervention in food markets was minimal. The Ethiopian Grain Corporation (EGC) was established in 1960 as a public corporation for maximizing the export of high-quality grains, but by 1975 it commanded only 5 percent of the grain trade (Lirenso 1983; Holmberg 1977). The end of share tenancy in 1975, however, had an immediate effect on the cereal market. Relieved of the obligation of paying rents and tributes to landlords, smallholders ate more of their production, causing the share of marketed surplus to fall. It is estimated that by 1977/78 it had dropped to around 11 percent of production (Ghose 1985). This decline in the quantity of grain made available to the market caused urban consumer prices to rise. Between 1974 and 1979, the price index for teff (taking 1967 as the base year) rose from 105 to 237. The rise for wheat was even more marked, rising from 117 in 1974 to 360 in 1979.

The Government's first response to rising urban prices was to establish the Agricultural Marketing Corporation in 1976. This agency absorbed the functions of the Ethiopian Grain Corporation (EGC) and began operations with a special focus on grain procurement for public distribution and on cereal price stabilization. Grain supplies for the towns and the army were secured via a quota system. In the late 1970s, Peasant Associations were obliged to deliver a minimum of 10000 kilogrammes of their annual produce to the AMC at fixed prices. In 1981, this quota was raised to 15000 kilogrammes, without a concomitant increase in producer prices.

Wholesale grain prices were officially fixed for many of the provincial markets by the end of 1976, establishing a pricing structure that remained essentially unchanged until the late 1980s (except for a 2.2 percent increase in the price of teff and sorghum in 1982/83, and further increases of 7 to 10 percent in 1988). Although AMC farm-gate prices were generally fixed for the country as a whole, they were substantially lower than market prices.

For example, the AMC price of teff in 1985 was 4.5 birr per kilogramme, compared with a retail price of 7.7 birr in (surplus) Gojjam and 15.7 in (deficit) Wollo. This suggests that households in deficit regions felt the effects of procurement relatively more than did those in surplus regions. As a result, several authors lay a large blame on the quota system for famine causation (Luling 1987; Winer 1989). The third point implies that all farmers, including the poor, have gained by price deregulation, although to varying degrees by region.

The greater burden on households in drought-prone regions may have been true in relative terms, but exceptions to the generalization deserve to be noted. Although it is often argued that "all farmers, regardless of the size of their harvest, had to meet their quota" (de Waal 1991), this was not the case everywhere. In 1988/89, only 18 percent of the 550 households interviewed for this study sold any of their harvest to the AMC, and the average quota delivered was only 2.7 kilogrammes per person. This represented just 3 percent of harvest, or less than one day's work on a food-for-work scheme (see Chapter 6).

Indeed, roughly 51 percent of the households interviewed never paid any quota at all since procurement began. Of those that did pay, there was a clear wealth bias in quota allocation. In 1988/89, poorest households paid an average of 100 grammes per person to AMC versus an average of 7 kilogrammes per capita paid by the relatively richer households. This evidence of careful quota allocation at a household level does not discount the likelihood that some farmers were forced to sell their stores, or even to purchase grain, in order to meet their obligations (Chole 1990; de Waal 1991). However, it does highlight the fact that quota-setting, as well as actual procurement, varied considerably in time and place.

Acting in tandem with the quota system (which did permit farmers to sell grain on the open market after procurement obligations had been met), was a policy that required licensed private grain traders to make at least 50 percent of their purchases available to the AMC, also at fixed prices. In the high-production regions, this requirement covered 100 percent of all privately purchased grain. The most important regions in this respect in 1986/87 were Shewa (which provided 31 percent of all cereal procurements at a national level), Gojjam (28 percent), and Arssi (20 percent) (Ethiopia, Central Statistics Authority 1987b). Private traders were paid 5 birr per 100 kilogrammes above the official farm-gate purchase price. Traders failing to fulfil their quota lost their license.

In order to control the operation of private traders, interregional trade of cereals (and even the movement of labour) was strictly regulated. In 1973, an estimated 90 percent of marketed grain was handled by some 20000–30000 private merchants (Holmberg 1977). However, between 1976 and 1980, the spreading influence of the AMC was paralleled by the declining influence of traders, constrained by legal prohibitions on small-scale commercial activities. Aside from the requirement of certified licenses (of which only 5000 were issued in 1986), road-blocks were erected at the entrances to all towns and large villages, aimed at controlling (and taxing) grain and population movements (de Waal 1991). Individual farmers were normally permitted to move 100 kilogrammes of grain through road-blocks so long as this did not involve crossing administrative boundaries.

However, transport restrictions within provinces were not applied everywhere with equal vigour. An example of the impact of policy restrictions during 1985/86 relates to UNICEF's cash-for-food programme (for more details see Chapter 6). Based on the distribution of cash in famine zones rather than food, this programme assumed that in certain areas it was a lack of purchasing power, rather than a lack of food, that was causing hardship. It was therefore expected that recipients of cash would tap into distant regional markets where food could still be obtained at a reasonable price.

At one site where this was tried, farmers travelled up to 100 kilometres to buy grain only to have it confiscated on the way home and be accused of illegal trading. Even by travelling at night and off the roads it was difficult to escape the militia enforcing this policy. Thus, most recipients were forced to stay home and contend with local inflationary effects of the cash transfer. A survey later established that due to these difficulties, 77 percent of the cash recipients would have preferred to be given food aid rather than cash (Webb 1989b). This provides a clear example of when a change in market policy was at least as essential as a change in grain prices.

At another, more remote, site where fewer militia were operating, 83 percent of participants were glad to receive cash rather than food, because food transportation to their distant location would have entailed delays and more hardship. Thus, if market integration is poor the effects of a crisis on local prices, or indeed on a relief project, may be quite different from one place to the next, making the task of price monitoring at more aggregate levels somewhat questionable.

Infrastructure Constraints

Policy restrictions were not the only constraint to market operations. A lack of adequate infrastructure also hinders market integration and a more equal sharing of scarcity among regions, and is largely responsible for the high margins characteristic of private sector marketing (Christensen 1991; World Bank 1993).

The dearth of rural infrastructure in Ethiopia has been widely commented upon (Iliffe 1987; Goyder and Goyder 1988; James 1989; WFP 1990). Ethiopia has only 90 centimetres of road per person, compared with 930 centimetres per person in Zimbabwe and 1230 centimetres in Botswana (von Braun, Teklu, and Webb 1991). What is more, within Ethiopia there remain wide disparities in coverage. The highest density of road coverage is found in the central and western grain-producing regions. Shewa has 3.14 kilometres of road per square kilometre, whereas Bale has only 0.11 kilometre per square kilometre (IFAD 1989). Thus, almost 90 percent of the country's population still lives more than 48 hours walk from a primary road (WFP 1989). It has been estimated that only 2 percent of villages in Wollo can be reached by all-

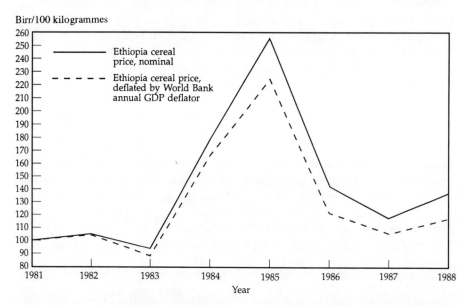

Figure 3.3. Cereal price index for Ethiopia, 1981 to 1988, based on nominal prices (i.e. not deseasonalized) (source: computations based on data from the Relief and Rehabilitation Commission; US$1 = 2.07 birr)

weather roads, making the movement of food to remote markets extremely difficult (OXFAM 1984).

It should also be pointed out that the national road system traces a radial pattern, with Addis Ababa as the hub and spokes leading to individual regions. The network connecting different regions, independently of Addis, is very underdeveloped. The role of such infrastructure deficiencies for market segmentation is addressed implicitly through more refined analysis below.

Links between Prices and Famine

There was a dramatic increase in both the nominal and the real price (inflation indexed) of cereals during 1984/85 and 1987/88. The national price index for cereals, computed as a weighted average across all provinces, increased from 100 in 1981 to over 220 in 1985 (Figure 3.3). It rose again, although not nearly so sharply, in 1988.

Although price formation is a complex process, which requires simultaneous consideration of supply and demand factors, it is clear that much of the increase can be explained in terms of production failure. A statistical regression modelling of changes in the real price index of cereals as a function of supply changes (that is, of domestic cereal production in the same year and

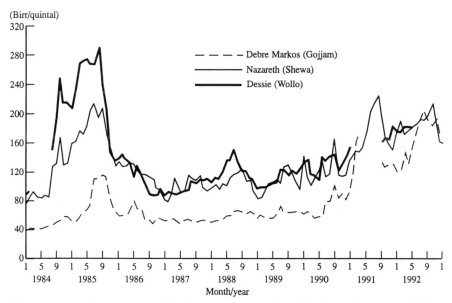

Figure 3.4. Market price of teff in the regional capitals of Gojjam, Shewa, and Wollo, 1984 to 1993, based on nominal prices (i.e. not deseasonalized) (source: constructed from data provided by the Relief and Rehabilitation Commission; US$1 = 2.07 birr; 1 Quintal = 100 kilogrammes)

of the previous year's ending food stocks) shows that a 10 percent decline in production results in a 14 percent increase in cereal prices.[9] Interestingly, a comparable national-level price response of 15 percent has been computed for Sudan (Teklu, von Braun, and Zaki 1991).

However, price increases observed for Ethiopia as a whole were not felt uniformly across the country. Figure 3.4 shows that in Dessie (provincial capital of food-deficit Wollo), the nominal price of teff between 1984 and 1989 was generally more than 100 percent higher than in Debre Markos (main market of Gojjam), which is a surplus-producing region enjoying stable production. During the 1984/85 famine, the Dessie price reached three times that of Debre Markos.

Although the picture is more mixed for less-favoured crops, such as wheat and sorghum, similar differences in absolute prices were observed between regions. On the one hand, crop prices were consistently higher in Dessie than in Debre Markos, with the price gap increasing during the major crises of 1984/85 and 1988. On the other hand, the Dessie price was lower than that of Nazareth (used here as the major market centre of Shewa), except during times of crisis, when the Dessie price leaped temporarily above that of Nazareth (Figure 3.5).

There were also consistent differences in absolute price levels for the major cereals within regions. In Wollo, the cheaper, more inferior cereal (sorghum)

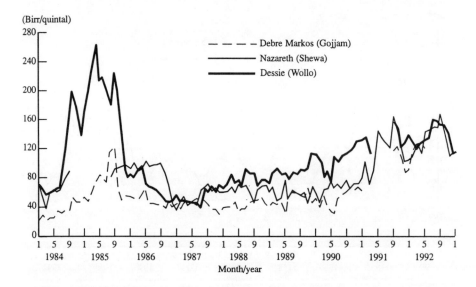

Figure 3.5. Market price of sorghum in the regional capitals of Gojjam, Shewa, and Wollo, 1984 to 1993, based on nominal prices (i.e. not deseasonalized) (source: constructed from data provided by the Relief and Rehabilitation Commission; US$1 = 2.07 birr)

mirrored price movements of the more expensive cereal (teff), but at a lower level. Sorghum came near the teff price only in early 1985, as demand for cheaper grain rose with declining income. In Debre Markos, demand for sorghum (usually relatively low in teff-surplus Gojjam) was so great that the sorghum price exceeded that of teff during August and September of 1985. Similar patterns occurred in each of the major cereal-producing regions of the country.

Although the general pattern of price movements during crises was similar in all regions, there were often time-lags. For example, sorghum prices reached their peak in Dessie five months before they peaked in Debre Markos. The same lag applied for wheat (Figure 3.5). This time difference indicates that the transmission of price signals is not fully efficient, even between neighbouring regions—a gap that is largely explained by the policy and infrastructure restrictions outlined above.

Market Disintegration

These examples of market dysfunction suggest that the monitoring of prices in one major market alone may not adequately reflect price movements in other parts of a region. It may therefore be important for famine early warning systems to consider not only prices in principal regional markets, such as Dessie or Harer, but also those of secondary market centres. For example, in his analysis of the 1973/74 famine in Wollo, Sen (1981) examined price data from Dessic (the provincial capital). Discovering little evidence of price increases, he concluded that there was no shortage of food at that time, only a shortage of purchasing power:

> A remarkable feature of the Wollo famine is that food prices in general rose very little, and people were dying of starvation even when food was selling at prices not very different from pre-drought levels (Sen 1981).

However, conditions in Dessie may not have been representative of conditions in other parts of the same region (see Cutler 1984; Baulch 1987; Iliffe 1987; McCann 1987b; Devereux 1988).

Time-series data on rural markets in Wollo during the 1970s are not available to settle this debate. On the other hand, as discussed above, the 1984/85 famine in Wollo took place with a reduction in food output and very large price increases. Despite similar price movements for teff across markets in Wollo during 1984–1989, a monitoring of Dessie prices alone would not have signalled the sharp rise (more than 100 percent) that occurred in mid-1984 in a secondary market, such as Alamata, during the worst period. Similarly, the steep rise in teff prices recorded in Motta (a secondary market in "surplus" Gojjam) from April to June 1985 would not have been spotted by monitoring prices only in the regional capital, Debre Markos.

There is no doubt that the famine of the mid-1980s was accompanied by dramatic price increases. The frequency of such crises causes an erosion of the purchasing power of the poor and does not permit rebuilding of the local economy. This development, if unchecked, can lead to famine situations with fewer and fewer obvious price signals. Thus, firm relationships of the past may not hold for the future. The longer term impact of price policy changes, including the abolition of quota procurements, fixed prices, and regional trade restrictions on production and prices, are as yet unknown. Indeed, the direct effects of the previous procurement policy on smallholder expectations and marketing behaviour was poorly understood (Lirenso 1987; Franzel, Colburn, and Degu 1989).

The collection of prices continues to be a mainstay of early warning systems, but questions about their message remain. Prices are not necessarily an "early" predictor of famine. Even in Wollo steep increases became apparent only in late 1984, long after the impending catastrophe had been signalled from agents in the field. Thus, a better understanding of market failure is crucial to early warning.

The present analysis has shown that short-term market disruptions and price explosions were commonplace. These did not always move synchronously because integration among regions is poor. Trade barriers (although not fully effective) and poorly developed infrastructure were to blame. As a result, local market integration between regional capitals and off-the-road markets was fairly strong. Although lagged, price developments in such markets were largely driven by developments in the subregion.

Although domestic food-market liberalization and integration in the international markets are keys to stimulated economic activity, regional specialization, and incentives for farmers, they need to be considered with caution under conditions of acute famine. Some time will be required to establish the institutions necessary for effective utilization of the free market: namely, active financial markets catering to traders and consumers; improved market information systems; and a strong supply response at farm level, facilitated by effective public and private input supply. Thus, resource- and income-poor households will remain vulnerable to sudden price rises associated with food shortages for some time to come.

The next chapter seeks to examine in more detail the types of household that are most vulnerable to price rises and food shortages, and what such households do in their attempts to mitigate the worst effects of constrained access to food. Private coping strategies are considered, as well as the production, income, and consumption effects of food crises on different households according to wealth status, gender of household head, and geographical location.

NOTES

1. The Central Statistical Authority of Ethiopia altered the way in which it collected crop production data in 1979. The calculations presented here use a correction

factor of 1.27 (which represents the difference between the base level of 1979 compared with 1978) to make the post-1979 data comparable with pre-1979 information. The following population growth rates were used: 1961–1974, 2.63; 1974–1979, 2.74; 1980–1984, 2.8; 1985–1989, 2.5; 1990–1992, 2.8.

2. The correlation between cereal production and cereal availability in Ethiopia is above 75 percent (Webb, von Braun, and Yohannes 1992).

3. During the 1980s, US$1 was officially exchanged for 2.07 birr. The rate was changed to 5.00 birr for US$1 in 1992.

4. Drought is not the only climatic factor that causes production failure in Ethiopia, frost, hail, and floods also play a part. However, drought is the most important climate-related determinant of cereal production at a national level.

5. The model specifications are kept simple because of a lack of data on fertilizer use and response and price response, which would be needed for more elaborate exercises. The model assumes that the effect of incremental rainfall on production decreases at the margin (to be picked up by a rainfall-squared variable). A dummy variable was included to separate the time-series data into two portions (pre- and post-1979) so as to compensate for the introduction of new methods for calculating farm yields and production.

6. For more details of the agricultural policies of the 1970s and 1980s see Göricke (1979, 1989), Ghose (1985), Griffin and Hay (1985), Rahmato (1987), IFAD (1989), and Webb, Zegeye, and Pandya-Lorch (1992).

7. For details of the agricultural system prior to 1975 see Pankhurst (1966), Hoben (1973), Cohen and Weintraub (1975), Cohen, Goldsmith, and Mellor (1976b), and McCann (1987a).

8. For details of the controversial villagization programme see Cohen and Isaksson (1987a), Luling (1987), Lirenso (1990), and de Waal (1991).

9. The regression model referred to has an R^2 (explanatory power) of 74 percent. For details see Webb, von Braun, and Yohannes (1992).

4   How do Households Cope?

What do people do when faced with the threat of starvation? In most cases, everything in their power. The people who die during famine should not be seen as passive victims but as losers of a hard-fought struggle for survival. Much can therefore be learned from that struggle, both about limits to private coping capacities and about how to improve public interventions.

Much has been written since the 1984/85 disasters about the potential for allowing traditional systems to deal in their own ways with food crises (Turton 1985; Hutchinson 1991; Wolde-Mariam 1991; Keller 1992). The implication is that public (usually external) famine interventions can be more disruptive than helpful, generating dependency on non-local resources and a disincentive to raise local productivity. There is little doubt that the vast majority of vulnerable households in Africa survive through famines without any outside help whatsoever. Consequently, a lot more needs to be known about indigenous support systems, local indicators of impending stress, and alternative sources of sustenance during crises.

However, there is a danger of idealizing the private coping capacity of poor households, and of establishing a false dichotomy between private and public spheres of action. Private, traditional initiatives are undoubtedly the key to survival; public interventions represent a short-term pain reliever at best. But, when food production, distribution, and consumption systems upon which private initiatives depend break down (as they have done repeatedly in Somalia, Mozambique, and Ethiopia), public intervention becomes vital to protect and stabilize the consumption entitlements of the poor. An understanding of how households are affected by, and respond to, famine is therefore crucial to the improved design of public interventions that can support, rather than inhibit, private action.

COPING OR SUFFERING?

The options for dealing with famine are widely referred to under a composite heading called "household coping mechanisms." The pattern of coping, largely determined by the pre-crisis characteristics of individual households, involves a succession of responses to increasingly severe conditions (Jodha 1975; Cutler and Stephenson 1984; Shipton 1990). This does not represent an overnight awakening to danger, rather a progressive narrowing of options that leads from broad attempts to minimize risk in the long term through actions designed to limit damage caused by a crisis, to extreme measures aimed at saving individual lives, even at the expense of household dissolution.

These various actions can be grouped for analytical purposes under three headings: risk minimization, risk absorption, and risk-taking. The first stage involves insuring against risk in an environment of limited credit and insurance markets. It involves measures of savings, investments, accumulation, and diversification.

The next stage of coping involves a draw-down of investments, calling in loans, and searching for new credit. As capital for investment dwindles, consumption of food and nonfood items becomes restricted, stores of food are drawn down, and the number and variety of potential income sources available become crucial to survival.

The last stage of coping, which may become inevitable if famine persists and food aid does not arrive, involves the collapse of normal systems of survival and the adoption of abnormal ones. At this point, the diet is dominated by unusual "famine foods" (roots, leaves, rodents), and households sell their last assets, including their fields, homes, and clothes. If they are still able to do so, some households break up and leave to search for assistance among distant relatives or at relief camps. If they are unable (or unwilling) to move, many individuals are forced into a passive conservation of energy (sitting waiting), which may ultimately lead to death.

This sequence of events shows that many of the actions taken to survive become increasingly irreversible as conditions get worse. Once a cow has been sold, it is no longer available for milk production. Once a field is sold, it cannot be cultivated by the former owner the following year. At the same time, households that progress along this continuum become increasingly vulnerable to a continuation of the crisis. Unless conditions change or external help arrives, each coping action at best delays the onset of the next stage.

Not all of the actions taken are beneficial, either to the household or to its environment. Reducing basic food intake to minimal levels or breaking up a family to enhance chances of survival of individual members entails much suffering. Indeed, there must come a point where the term "coping" becomes more of an aphorism than a useful description. Desperate actions, such as the cultivation of marginal land or wholesale felling of trees to sell as firewood, have serious consequences for future environmental development and income generation. The sale of children or the abandonment of elderly relatives is a sad way to have to "cope". As Davies (1993) points out, once prevailing rule systems have collapsed under stress, "adapting" may be a more appropriate term than "coping".

The following sections examine elements of the sequence of actions aimed at surviving famine. No attempt is made here to specify the precise timing of events in the sequence, an issue that has preoccupied some researchers (Watts 1988; Frankenberger 1991). Household responses involve trade-offs between, and within, various coping options. In other words, different households within a community stand at different points along the continuum, and their response to threat varies according to their endowment base, access to community support, and access to public interventions. The focus here is therefore

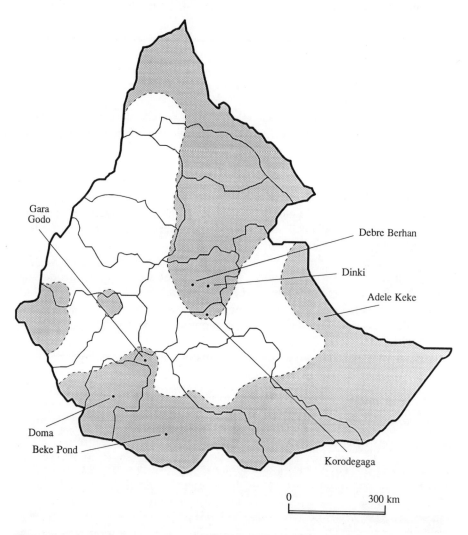

Figure 4.1. A map of survey sites in Ethiopia. The shaded areas represent the zones most affected by famine and drought during the 1980s. Debre Berhan, Adele Keke, and Gara Godo are highland sites. Doma, Korodegaga, Dinki, and Beke Pond are in the lowlands

Table 4.1. Profile of survey communities, 1989/90 (source: survey data, 1989/90)

Community surveyed	Admini- strative region[a]	Number of households in village	Number of households surveyed	Mean household size	Main ethnic groups
Doma	Gamo Gofa	230	100	6.2	Gamo
Korodegaga	Arssi	186	102	6.3	Oromo
Dinki	Shewa	120	57	6.4	Amhara/ Argobba
Gara Godo	Wolayta[b]	5180	65	7.9	Wolayta
Beke	Sidamo	106	48	6.2	Borana/ Gabbra
Adele Keke	Hararghe	350	107	5.8	Adere
Debre Berhan	Shewa	746	68	6.5	Amhara/ Oromo

[a] Pre-1987 regional names used (see Chapter 2).

[b] Although Wolayta is not strictly an administrative region (it being part of Sidamo Province), it is commonly referred to as if it were.

on the specifics of adjustments without implying that there exists an absolute chronology for such responses.

THE SURVEY SETTINGS

Questionnaire surveys were conducted in seven parts of the country (Figure 4.1). Eight of Ethiopia's 85 ethnic groups are represented in the sample, as are three major religious groups: Orthodox Christianity, Islam, and animism (Table 4.1). And three distinct farming systems are included: plough-based annual cultivation; hoe-based mixed farming (annual plus perennial crops), and pastoralism.

However, the sites were not chosen at random, nor were they designed to be "representative" of the country as a whole. For either of the latter two conditions to be met, a massive survey covering all provinces would have been required. Instead, given the nature of gaps in our understanding of famine, the present survey placed greater emphasis on depth than on coverage.

The seven sites selected met two required conditions. First, each location suffered problems of food supply and/or food availability between 1984 and 1989 (but hardships that were not caused by direct military disruption of production). Indications of crisis at a local level were found in district-level government famine early warning reports, donor monitoring activities, and periodic nongovernmental organization (NGO) situation reports.

Second, each area received external assistance during the crisis. We wanted to understand how households coped in their own manner, and how relief interventions were able to contribute to private coping strategies. Prior to site selection, 10 months were spent in Ethiopia reviewing project documents and donor reports, interacting with local university researchers, and visiting field locations in order to identify relief and rehabilitation projects that kept good records of their own activities, costs, and impact. Seven projects were retained for the study, and the agencies behind these projects collaborated generously with this study at every stage.

A brief description of the communities is appropriate here. The first village is called Doma. This village was founded in 1985 with support from the Relief and Rehabilitation Commission (RRC). The settlers chose to leave overpopulated villages in the highlands, which had been badly hit by drought in 1983/84.[1] They moved to an area of virgin forest at the foot of the mountains, west of Arba Minch. Although served by two sizeable rivers, providing limited irrigation potential, the area was found to be semi-arid. Drought destroyed the 1985 harvest. Given the hardships of relocation, low food supplies, and malaria (to which highlanders had not previously been exposed), a crisis was not long in developing. A feeding camp was set up in Doma by RRC and UNICEF in early 1986. This closed later that year when the crisis abated. However, similar harvest failures in 1988, 1989, and 1990 threatened to trigger a famine greater than that of 1985.

The second site, Korodegaga, also faced a renewed crisis in 1989/90 following its earlier experiences in 1985/86. This village lies on the floor of the Rift Valley in the drought-belt of an otherwise prosperous region.[2] Continued droughts since 1981 resulted in severe food shortages in 1986. Several respondents noted that they were afraid at that time "of becoming a second Wollo."

The third site is in the foothills of the Rift Valley on the border between Shewa and Hararghe. Unattainable by motor vehicle until June 1986 (when a Food-For-Work project built a road), this village could be reached from the highland town of Debre Berhan only by a two-day mule ride down steep mountain slopes. Food shortages were severe in 1985, and a cholera epidemic caused high mortality.

Gara Godo is located in the densely populated region of Wolayta (Rahmato 1990a). Mean population density for Wolayta stood at 222 persons per square kilometre in 1988, compared with a national average of 40 persons per square kilometre (Redd Barna records). Table 4.1 shows that the mean size of households surveyed was indeed higher than at the other sites—7.9 compared with an average for the other sites of 6.2. Also, the number of households per village is higher than in other areas. Farms are small and households depend heavily on root crops, such as enset. A lack of rain in 1983 and 1984, coupled with bacterial disease that destroyed much of the enset crop, led to famine here before most other parts of the country: the so-called

"Green Famine." As Goyder and Goyder (1988) commented: "To the casual observer, [Wolayta], after the main rains had begun in July 1984, looked like a verdant and fertile area; yet people starved in their own homes."

Beke Pond is not a village but a watering hole. Semi-nomadic pastoralists settle in camps around this pond for up to 7 years in a stretch. The inhabitants of three camps were interviewed: one inhabited by Gabbra camel-herders, and two inhabited by Borana cattle-herders. Both pastoral groups were affected by the collapse of livestock prices relative to cereal prices in 1985/86. An RRC feeding camp in nearby Mega was still occupied by pastoral families in 1987.

The last two sites, Adele Keke and Debre Berhan, are both located in the highlands. Adele Keke suffered a crisis when drought spread eastward from Wollo in 1985/86. Debre Berhan also faced a crisis later than the lowland villages, when shortfalls in food and fodder in 1986/87 (caused in this instance by frost damage and hail storms) resulted in livestock mortality, asset depletion, and, finally, food shortages among the poor. The survey focused not on the town of Debre Berhan, but on four surrounding villages called Fagi, Karafino, Kormargefia, and Milki.

Each of these sites did not feel the worst of the famine at the same time. The period of famine duration and the crisis peaks most often referred to by respondents are indicated in Figure 4.2. The figure shows that although all

	Previous Worst Famine	1983	1984	1985	1986	1987	1988	1989	1990
Adele Keke	1981								
Beke	1978								
Debre Berhan	1957								
Dinki	1957								
Doma	1978								
Gara Godo	1974								
Korodegaga	1981								

Figure 4.2. Duration and peaks of hunger period as reported by sample households, by sample community. Crisis peaks are indicated by bolder lines (source: survey data, 1989/90)
[a] Doma households left their old villages in the highlands in early 1985 to found the new village in the lowlands

sites experienced a crisis of varying severity at some stage between 1983 and 1988, the peaks did not always coincide; some were worse off during 1985, others not until 1988.

Methodology

At each site, a list of households that participated in relief interventions was drawn up from project records. A random sample of households was drawn from these lists to make up half of the site-specific sample. The other half (a control group) was drawn at random from each Village Council's own list of community members.

Given the differences in sample size at the survey sites, a system of weighting is used whenever data are pooled across sites. The weights are derived from estimates of the number of households in each of the districts in which the surveys were undertaken. These estimates are based on the most recent population census information available.

The variety of data available permits analysis at three levels of disaggregation: by agro-ecology, wealth status, and gender. Such disaggregation allows us to view the problem from many angles. It also avoids hiding diversity behind a mask of averages. Where location is used as a stratifying factor, the sites are divided into three groups: highland, lowland, and pastoral.[3] Three of the communities are located in highland areas at more than 1500 metres above sea level and receiving an average of more than 1200 millimetres of rainfall per year (Adele Keke, Debre Berhan, and Gara Godo). The other four sites are located in the lowlands at altitudes of less than 1500 metres above sea level, and receiving an average of less than 700 millimetres of rain per year (Doma, Dinki, and Korodegaga). The fourth lowland site is the semi-nomadic, pastoral community (Beke Pond).

Where wealth is used as a stratifier, three equal "income groups" are used to represent differing levels of wealth. These groups were not pre-identified for the survey. They are based on the post-survey calculation of total net income per person for surveyed households based on the 1988/89 season. All income sources were considered in that calculation, including the value of home-produced crops and livestock. Households were then ranked into three equal groups according to their position on the local income scale. Given the real poverty of all households in Ethiopia in absolute terms, those at the very top of the income scale were still poorer than the poorest households in most other African countries. Thus, the three groups are distinguished here not as rich or poor, but as "poor", "poorer", and "very poor".

Finally, some data are reported using gender of the head of household as a stratifier. This can be important because female-headed households are often smaller, poorer, and more vulnerable than those headed by men—not *because* of the gender of the head, but because of what the gender of head signals in terms of other household characteristics. Where the head is defined as female in this text, this constitutes a *de facto* headship; that is, there have been no adult males present in the household for at least six months. Korodegaga had the highest share of woman-headed households, with 24 percent. The Adele Keke and Gara Godo sites each included 11 percent of households headed by a woman. The share at each of the other four sites was less than 10 percent.

In the following sections household data are examined both within and across sites according to these three stratifiers. Following the stages of crisis response outlined above, the analysis focuses on risk minimization (ex ante) and risk absorption (ex post) responses in relation to production, asset holdings, income, and food consumption. This is followed by a brief consideration of the final (catastrophic) stage of risk-taking, when households generally migrate or collapse.

RISK MINIMIZATION IN A RISKY ENVIRONMENT

Ethiopian agriculture is characterized by low technology, low productivity, and high risk. Yields and crop output are poor even in years of relatively good rainfall. In 1988/89 (a good year), the average cereal yield obtained by the 547 sample households was only 740 kilogrammes per hectare in the highlands and 300 kilogrammes per hectare at lowland sites. The highland figures match the 700 to 800 kilogrammes per hectare reported for highlands in Eritrea and Wollo (Collinson 1987; Wolde-Mariam 1991). These yields translate into an average of 111 kilogrammes of cereals produced per person in this "good year" in the highlands, and only 50 kilogrammes per person in the more arid lowlands.

Farmers offer four main reasons for their low productivity. In the highland villages, the first issue raised was the poor quality of the land. The average farm size held was only 0.15 hectares per person. This small area is affected by soil erosion, as well as by problems such as stoniness and waterlogging. At Debre Berhan, 35 percent of households reported very severe problems of stones in the ground that inhibit ploughing, and 48 percent reported very severe waterlogging when the rains arrive. At lowland Dinki, waterlogging was less of a problem, but stoniness was very severe for over 50 percent of households. Pulling stones (large and small) out of the soil is a labour-intensive and never-ending task. Inhibiting soil erosion or attempting to assist water percolation through soil bunding and terracing technologies are also highly labour-intensive, and such actions tend to reduce the potential cultivated area still further.

The second major constraint to productivity is a lack of livestock. Constraints include the inability to ensure timely and effective ploughing because of a lack of oxen, low income security because of a lack of milking cows and a lack of transport animals to carry produce to market, and low animal productivity (of meat, milk and calves) because of diseases and a lack of adequate grazing throughout the year. Without animals, most farmers (let alone pastoralists) would be poorer and more vulnerable than they already are (Wolde-Mariam 1991).

Thirdly, there were reported problems with seed stock. Most farmers sampled stated that their seed stock of staple cereals, tubers (enset), and perennials, such as coffee and chat (*Catha edulis*, a perennial bush that produces narcotic leaves), is very poor. Germination rates are low, ability to compete with weeds is low, and pests and diseases often devastate otherwise good crops. Hybrid cultivars aimed at higher productivity or drought resistance remain scarce in most rural areas. As a result, the purchase of seeds at time of planting, either because the previous harvest was too small to permit seed storage or because germination rates are so low, becomes an onerous expense. Less poor households spent an average of 21 birr on seeds for the 1988/89 season, compared with 16 birr on other purchased inputs. By contrast, the very poorest households (most of them in the lowland sites) spent an average of 81 birr on seeds and nothing at all on any other inputs.

Aside from seeds, the lack of other farm inputs constitutes a fourth major factor constraining agriculture. As noted in Chapter 3, physical access to markets is a problem everywhere, but even where markets are accessible the availability of fertilizers, pesticides, credit, improved tools and ploughs, veterinary services, and even extension advice is severely limited. For example, only 1 percent of the sample households (all in the highland communities) used any chemical fertilizer during the 1988/89 season, and less than 15 percent hired any labour. None had ever opened a bank account, and only 17 percent reported membership in community-based savings societies such as the *iqqub* or *iddir*—again, these were relatively richer highland households (Aredo 1993). These constraints to investment, and to credit that might stabilize fluctuations in food intake, serve to restrict potential productivity gains in agriculture (Winer 1989; Stahl 1990; Lycett 1992).[4]

Given such harsh "normal" conditions, households attempt to cope with climatic or economic shocks by spreading risks—both on the farm and off the farm. Many of the risk-spreading measures used in Ethiopia are commonly found across Africa (Cekan 1990; Shipton 1990; OFDA 1991b).

Agricultural Diversification

To spread risk across the farm, many cultivators grow a range of crops and of species, including food and cash crops, early-planted as well as late-planted varieties. Many of these plants are intercropped to reduce the risk of total failure in any one crop. Thus, sorghum is planted between chat in Hararghe, and barley is mixed with lentils and wheat in northern Shewa. Crop mixes are also driven by the desire to follow a flexible production schedule that adjusts to variations in rainfall patterns. Staggered planting is common, with crops of different maturing periods being planted successively rather than simultaneously. This reduces labour conflicts and protects against total seed loss due to mid-season droughts.

Second, high variability in rainfall encourages farmers to disperse their individual holdings. In the dissected terrain of Ethiopia's Rift Valley foothills a dispersal of field plots by altitude (different points on a slope) is favoured so that variations in microclimate can be capitalized on. The likelihood of total crop failure in a set of dispersed fields is lower than in a group of contiguous fields.

Diversification and dispersal are also the most common methods used to protect livestock holdings. Diversification involves the husbandry of a mixed animal stock; that is, both small and large ruminants as well as browsers. This permits a better utilization of feed resources (because of more varied demand) and easier disposal of small stock before unproductive large stock during times of crisis. Dispersal involves spreading the risk of herd loss by splitting the herd into semi-autonomous groups kept in the same region or arranging for stock to be husbanded by relatives or contracted labour in more distant locations.

Third, some farmers overseed favoured parts of their fields in order to maximize the chance of plant survival. They may subsequently practice selective weeding—only assisting the most favourable shoots thereby minimizing labour lost on unproductive plants.

Fourth, farmers in the most risky environments rely increasingly on shorter maturating plant species in the search for protection against drought. This increasing emphasis on short maturing crops does not necessarily mean improved resistance to drought. Some improved varieties of maize mature in 90 days compared with 120 days for most traditional varieties, but this does not always reduce their resistance to moisture deficits. The attraction of short maturation is associated more with the ability to end the "hungry" season as early as possible (thereby reducing human physiological stresses), and the potential in some areas for planting a second crop in one season (for example, barley following short-maize or lentils in highland Ethiopia).

Diversification off the Farm

Given the constant threat of crop failure, farmers adapt by seeking income from sources other than agriculture. Nonfarm sources have been found to provide an average of 38 percent of total income in rural African households (von Braun and Pandya-Lorch 1991). For example, Gambian households derive 23 percent of total income from nonfarm activities (von Braun, Puetz, and Webb 1989). In Burkina Faso, the share of total income obtained from nonfarm sources is roughly 30 percent (Reardon, Delgado, and Matlon 1992).

In Kenya and Rwanda, the share of nonfarm income represents 40 and 60 percent, respectively (Kennedy and Cogill 1987; von Braun, de Haen, and Blanken 1991).

The income levels for each of the survey sites in Ethiopia are given in Table 4.2, reported for the top and bottom income groups for each survey site. The table shows that although a few relatively wealthy households were found in each of these communities, income for all households was extremely low in absolute terms. Total net income (calculated as income from agriculture and nonagricultural sources, minus costs), stood at only US$41.50 per person per year. This ranged from only US$10 per person among the very poorest households to US$131 in relatively less poor households.

Although some underreporting of income is inevitable (for example, income from any *belg* season harvest is excluded from these calculations), these figures give a valid indication of the degree of poverty experienced at the survey sites. For example, these dismal income levels compare unfavourably even with those of households in other famine-prone countries. In Sudan's province of Northern Kordofan, households in the lowest income groups of a

Table 4.2. Net annual income (mean net income per capita, US$)[a] per person for sample households at each survey site, by agro-ecological zone and income group, 1989/90 (source: survey data, 1989/90)

Survey site	Income groups[b]			Sample mean
	Poorest	Middle	Less poor	
Highland				
Adele Keke	14	63	140	73
Debre Berham	44	95	183	107
Gara Godo	4	16	37	19
Lowland				
Dinki	24	47	111	60
Doma	3	11	38	17
Korodegaga	–4	13	58	22
Pastoral				
Beke Pond	3	18	136	52

[a] Converted at US$1 = 2.07 Ethiopian birr.

[b] The income groups were calculated after data had been collected, not before. The groups are based on net annual income per household for 1989/90. Income includes: (i) own agricultural production (main season crop production, sale of livestock and its products, and income from farm labour); (ii) nonfarm labour income (received as wages for nonagricultural work, and income from artisanal activity); (iii) transfers (remittances, loans, gifts, dowry, inheritance, sale of food aid, in-kind income derived from food-for-work); and (iv) sales (income derived from selling own cash crops [cotton, sugarcane, coffee, chat], collected fuel products [wood, dung, incense], collected consumables [bush foods], and processed foods and drinks). The calculation of costs includes the purchase of seed, fertilizer, manual labour, and the rental of oxen and tools. Family labour is not costed for this calculation.

survey conducted in 1988/89 were found to have incomes of US$125 per person per year (Teklu, von Braun, and Zaki 1991). Similarly, a survey in Niger in 1991 showed that the poorest households at three rural sites obtained a yearly income of roughly US$300 per person (Webb 1992).

Although Ethiopian households are extremely poor in absolute terms, some are even poorer than others. Table 4.2 shows that Doma and Gara Godo have a narrower range of income and a much lower ceiling than do Beke Pond and Debre Berhan. The poorest households in Debre Berhan are, on average, wealthier than the richest households in Doma and Gara Godo. What is more, the poorest households in Korodegaga showed a negative net income as a result of high production costs not recovered in 1988/89 because of poor or failed harvests.

Such differences in income by site are due largely to variations in agricultural potential and nonfarm income-earning opportunities by location. The two sites with the highest incomes (Debre Berhan and Adele Keke) are situated in the highlands, with better crop potential and infrastructure than the lowland sites. Households in the highlands also have more opportunities for nonfarm employment; for example, labouring, trading in fuel products, or craft work.

Table 4.3 provides a breakdown of household income sources for the Ethiopia sample compared with households in Sudan and Niger. This table shows that dependency on agricultural income is high in all three countries (always at least 40 percent of total), but highest of all in Ethiopia, where it represents between 61 and 66 percent of total net income according to the income group.

Yet, even in Ethiopia, other sources of income are clearly important. The sale of fuel products, craft work, and other activities not related directly to the

Table 4.3. Sources of income (percentage of total net income per capita) for sample households in Ethiopia, Sudan and Niger, by income group (sources: survey data, 1988/89; Teklu, von Braun, and Zaki 1991; Webb 1992)

	Ethiopia		Sudan		Niger	
Income sources	Poor	Very poor	Poor	Very poor	Poor	Very poor
Cropping	42	55	30	25	43	40
Livestock	24	6	13	15	3	4
Wages	13	2	8	13	21	8
Fuel products	15	28	11	21	7	13
Handcrafts	1	3	2	5	1	2
Transfers	4	4	33	19	11	15
Commerce	1	1	3	1	14	18
Other	0	1	0	1	0	0
Total	100	100	100	100	100	100

farm accounted for 57 percent of total net income for Ethiopian households in the top income group, and 71 percent of income for the very poorest households. In Sudan and Niger, transfers, wage employment, commerce, and the sale of collected fuel products contribute the most to nonfarm income.

Within Ethiopia itself, the importance of nonfarm income varies by region, by season, and by gender. For example, pastoralists have little tradition of manual labour, but the sale of craft work (woven baskets) is of growing importance. Table 4.4 presents income source data for Ethiopia grouped by agro-ecological zone. As would be expected, income from cropping is of much less importance to the Beke Pond pastoral communities. Conversely, the sale of animals and animal products is hugely important, representing 52 percent of total net income. Craft work and wage income are the next most important activities.

There are also important differences in the income breakdown between the two groups of settled farm communities. For example, in the highlands the sale of animal products and nonfarm employment together contribute almost 18 percent of total income. In the lowlands, these two sources provide less than 2 percent.

By far the most significant non-cropping activity among lowland farmers is the sale of fuel products, such as firewood, charcoal, and dung, representing more than 35 percent of total income. This mining of natural resources, such as wood and potentially useful organic fertilizer, underlines the poverty of the lowland farmers. They are stripping their asset base because of short-term need. However, the danger of increased poverty when such resources disappear is great. The women of Korodegaga already travel 20 kilometres on a round trip from home to "forest" to market and back home. The stretch from

Table 4.4. Sources of income (percentage of total net income per capita) for sample households by agro-ecological zone (source: survey data, 1989/90)

Income source	Agro-ecological zone[a]		
	Highland	Lowland	Pastoral
Cropping	54	47	20
Livestock	18	9	52
Wages	11	1	13
Fuel products	5	35	0
Handicrafts	3	1	12
Transfers	6	4	2
Commerce	2	1	0
Rental income	1	2	1
Total	100	100	100

[a] The highland zone includes Adele Keke, Debre Berhan, and Gara Godo. The lowland zone comprises Dinki, Doma, and Korodegaga. The pastoral zone is represented by Beke Pond.

forest to market is travelled with a 30 kilogramme load of wood carried on the back.

Unfortunately, the market for many products (including fuel wood) cannot be guaranteed. Some markets are seasonal. Wages for farm labour and rental income from renting land or draft animals can be secured only during the rainy season. Craft work and nonfarm employment assume greater significance in the dry season when farm activity is low. Income from certain service activities, such as milling grain by waterwheel (a service provided by two households in Dinki), is also seasonal, because the busiest months are those following harvest.

Other markets are compressed during a crisis. Droughts or other crises tend to distort markets for both agricultural and nonfarm products and services. Drought tends to reduce demand for nonessential foods and fuel products because cash is conserved for the purchase of staple foods. This sharply reduces the earning options for many people, but particularly for women. For example, in a relatively good rainfall year (1989/90), women in 21 percent of the survey households were earning income from the sale of fuel products. In another 8 percent of households women were dependent on the sale of processed food and drink. However, during the mid-1980s famine less than 1 percent of all households were engaged in such activities. Instead, women were forced to sell their last asset of value (labour) alongside the men.

Similarly, in 1988/89, wage labour was a major source of income for men in 17 percent of households and for women in only 3 percent of the households. Women were more active than men in trade and in selling fuel products and food and drinks (local beers and spirits, roasted barley, and maize). Men were more active as manual labourers, but during the worst famine year the percentage of households in which both women and men were working as labourers rose to 25 percent. This despite a collapse in wage levels. In Dinki, for example, the wage for manual field labour fell from US$2 per day in 1983 to US$0.25 in 1985.

These income adjustments among households in Ethiopia indicate that the higher their dependency on agriculture for income (which implies limited access to alternative income sources), the greater the risk of income failure. Thus, households in drier areas, especially asset-poor households headed by women, are particularly at risk.

Strengthening Savings

Savings can take the form of food stores, real-value stores (household goods of value), and cash stores (Swift 1989). Cash stores are difficult to measure because people are always reluctant to share such information. No household admitted to holding a formal bank account, and few belonged to local savings groups.

Food stores are easier to verify. Most households store a portion of their harvest to tide them over until the following year, but only 35 percent of the 547 sample households had grain (food and seed) held in storage in 1989. Over 65 percent of households had nothing in store at all. The amounts stored were small, averaging only 14 percent of the previous harvest across all sites. In real terms, this translated into an average grain store of only 17 kilogrammes per person.

However, the ability to store food was not evenly distributed. The proportion of households with at least some grain in store ranged from 54 percent among less poor households (in the top income group), to less than 2 percent among the very poorest households. During the famine years, the number of households with empty stores rose dramatically.

As for assets, the most valuable (after livestock) tend to be housing materials (metal roofs and wooden posts), metal-framed beds, new clothes, and ploughs and ploughing harnesses. Nevertheless, in 1989 the mean value of assets still held by households that had survived five or more years of crisis was only US$66 per person. This ranged from US$114 per person among less poor households to only US$52 in the very poorest households. In other words, asset stripping for survival is only a limited contingency. Having survived a crisis, rebuilding the asset stock must be a high priority.

RISK ABSORPTION

Measures adopted by households to minimize risk are effective for only limited periods of time. Successive years of below average or poorly distributed rainfall have a negative effect on production, and hence on income and consumption. Such was the case not only in Ethiopia but also in other drought-prone countries, such as Sudan (Teklu, von Braun and Zaki 1991), Kenya (Kennedy 1992), Chad (IUCN 1989), and Burkina Faso (Webb and Reardon 1992).

Drought Impact on Crop Production

The crop response to inclement weather in Ethiopia was marked. Main season cereal yields across the survey sites in 1984/85 were very low, at 181 kilogrammes per hectare compared with 508 kilogrammes per hectare in 1988/89, but the impact of drought was not felt uniformly; there were large differences both by income groups and by agro-ecological zone.

For example, in 1984/85, the average output of sample households dropped to only 70 kilogrammes per person in the highlands, compared with less than 4 kilogrammes per person in the lowlands (Table 4.5). However, higher income households in the highlands maintained their average output close to 100 kilogrammes per person, whereas poorest households fell to only 43 kilo-

Table 4.5. Drought impact on planted area, cereal yields, and cereal output among Ethiopian households, by income group[a] (source: survey data, 1989/90)

	Highlands		Lowlands	
	1984/85	1988/89	1984/85	1988/89
Cereal yields (kg/ha)				
Poorest households	389	553	0	230
Less poor households	587	905	34	347
Cereal output (kg/person)				
Poorest households	43	47	0	13
Less poor households	95	176	7	88

[a] Main (*meher*) season only.

grammes per person. The advantage of owning oxen, being able to hire labour, and to purchase seed resulted in higher cereal yields (roughly three times greater) and output (more than double even in the worst year) than obtained by the poorest households.

The difference between income groups in the lowlands was less marked, mainly because there was almost total crop failure for all households considered. Table 4.5 shows that in 1988/89 the yield and output differential between highest and lowest income groups was narrower than that between income groups in the highlands. During the drought this differential narrowed further leaving no one with more than 7 kilogrammes in hand.

The main response to such drastic yield and output failures among smallholders appears to have been not a heavier investment in agriculture to make up the shortfall, but a concentration of scarce inputs on fewer fields and a search for alternative sources of income. This is demonstrated by a decline in area cultivated during the worst years. A lack of inputs led to a retrenchment of activities during the crisis years, rather than an expansion of area. The average area cultivated by sample households in 1984/85 (0.4 hectare per household) was only two-thirds of that farmed in 1988/89 (0.6 hectare).

However, there was a disproportionate decline among the very poorest farmers. Households in the top income group were cultivating 0.15 hectares per person in 1984/85, which rose to almost 0.2 hectares per person when the rains improved in 1988/89. The very poorest households, on the other hand, farmed only 0.03 hectares per person during the worst of the drought, compared with a mere 0.09 in the better rainfall years. The result of this area retrenchment, coupled with the yield collapse, was a 1985 main season cereal harvest at less than half the level attained after good rains in 1988/89.

It should also be pointed out that the short rains *belg* harvest of the 1985 season was a total failure at the one survey site where it might have made a difference, namely in Gara Godo. Three survey sites pursue planting during

the early rains (Debre Berhan, Dinki, and Gara Godo), but only in the latter does the *belg* harvest represent more than 5 percent of total annual harvest. In 1988/89 (a good rainfall year), the *belg* contributed 26 percent of total annual cereal production. This was most important for the very poorest households, for whom it represented 34 percent of total cereal production. Among less poor households the share was closer to 21 percent. However, in 1984/85, Gara Godo's small rains did not materialize and no crops were harvested.

Impact of Drought on Livestock

Drought and famine does not only compromise crop cultivation; they take a heavy toll on livestock. Despite extreme measures taken by owners to preserve their herds during the 1980s, the capacity of most smallholders to feed their animals was increasingly constrained by their declining income.

Some extraordinary measures were taken by households to keep their most valued animals alive. In Korodegaga, for example, three households (all headed by widows) shared all food available to themselves with their animals, particularly the cows and calves. This was seen as a logical attempt to safeguard the households' long-term survival. In Dinki, 69 percent of households sampled fed thatch off their own roofs to their oxen and milch cows. Cactus stems and fruit were used by 24 percent of households, and 5 percent collected creepers from the trees.

Despite such extreme preservation measures, loss of production and animal mortality during the famine years were high. In highland Debre Berhan, milk output was affected by reduced forage and fodder availability. The best milk yield from local cows held by sample households during 1989 was 2.2 litres per day. However, according to survey recall data, average output fell to 930 millilitres per day during the worst of the food crisis, with large numbers of cows drying up altogether. Similar shortfalls were experienced, but on a much larger scale, by the pastoral households at Beke Pond (see Chapter 5).

The ultimate production loss for livestock owners is death of the animal. Total livestock mortality for sample households as expressed in tropical livestock units (TLU) during the worst year was, on average, double the mortality experienced in 1989/90: 0.37 TLU per person compared with 0.17. In other words, crisis conditions had a severe impact on total animal numbers, which forced most owners to accept some losses in order to keep some animals alive.

The poorest households (with few animals) took some of the most extreme measures to keep animals alive, and were in large measure fairly successful. For example, households at the top of the income scale lost more livestock units per person than did the poorest households—0.23 TLU versus 0.13 TLU, respectively. This represented a loss of 88 percent of total herd stock for the higher income households versus a loss of 23 percent among the very poorest households.

On the other hand, even if relatively richer households lost more animals, they still survived the famine years with more stock in hand than the poorest households. In 1989, the poorest owned an average of only 0.17 TLU per person, whereas less poor households still held 0.42 TLU per person. This differential ownership of livestock after a crisis according to wealth has important policy implications for the rehabilitation of farming systems devastated by drought. If the poor are not successfully targeted by restocking projects, the goal of increasing food security and stabilizing incomes among the most vulnerable will not be easy to achieve.

Asset Sales

With crop and livestock output compromised, food stores depleted or empty, and most capital reserves exhausted, patterns of normal-year activity begin to change. As the search for income assumes a new urgency, asset sales become more common, new types of nonfarm work are adopted, and economic debts and social obligations are called in. Access to credit to stabilize consumption and to limit the distress sale of assets is crucial at this stage for a quick recovery after the crisis. If one can survive through the worst months by obtaining food (or other resources) on credit without having to sell assets, the post-crisis period will be much easier.

Wealthier households handle this stage of coping better than poorer households because they have more assets (equipment, durables, livestock) to part with, and they generally have better access to credit and other support systems. As expected, the poorest households had few assets to sell, and those that were sold tended to be of lower value. An average of 42 percent of the very poorest households sold some form of asset during the worst year of famine (mainly livestock, but also farm and household assets), compared with 64 percent of households in the higher income group. The value of income gained from asset sales during the famine by the latter households was equivalent to US$15 versus only US$3 earned by the very poorest.

There were also differences in the volume of assets sold according to the intensity of the crisis by survey site. For example, in Debre Berhan, where the crisis was relatively less severe than at other sites, only 19 percent of respondents sold any household items, such as tables, pots, and blankets. In Doma, by contrast, where famine conditions were severe, many people sold their own clothing (coats, dresses, shoes) and essential cooking utensils (dishes, cups, jugs). A number of particularly distressing cases were found in Doma. A former carpenter, forced to sell tools during 1986, could only find work in 1989 as a manual labourer. One respondent recalled that in 1986, "the local market was empty. The only people there were trying to sell a shirt or their own trousers."

Sales of productive assets represent the later stages of hardship. Because private ownership of land was prevented by law, land did not often change hands during the crisis. However, 28 percent of households sold at least some farm equipment and 56 percent sold livestock. Disposable farm equipment took the form of ploughs, sickles, harvest sacks, and rope. In Debre Berhan, no households sold any of their productive farm assets, but in Doma, 48 percent of the households sampled sold farm-related assets. Most of the latter were among the very poorest households.

Where livestock are concerned, most animals sold were male cattle, calves, and small ruminants. Nevertheless, draft oxen, cows, and donkeys (the principal mode of transport and haulage) were also sold as conditions worsened. In 1984, less than 1 percent of the households sampled sold any oxen or productive cows in order to purchase food. The following year, however, distress sales rose steeply, with 9 percent of households selling cows and 13 percent selling oxen. One man in Gara Godo sold his only ox for US$5 in 1985. When asked why he did so his reply was, "Forget the ox. If someone had offered to buy my son at that time I would have sold him."

By 1988/89 conditions had improved and only 3 percent were still selling oxen for food. But the long-term implications of such asset stripping are considerable. Fewer oxen (and ploughs) are available for the next farm season, income from animal products disappears, and sales of fuel products suffer because of transport constraints. These examples signal that rural capital markets do not function efficiently for the poor; that is, the poor had no means of protecting their assets and their productive efficiency. This underlines the relatively greater problem facing poorer households in attempting to reestablish post-drought production. There is little doubt that famine-related asset losses strongly impair post-crisis economic recovery and entrench food insecurity.

Community Support

There is a large anthropological literature underpinning the unresolved debate over the nature and extent of sharing within communities during times of stress (Turnbull 1972; Dirks 1980). Part of this debate centres on whether communities, or even relatives, share or do not share what they have during famines. Evidence on community exchange and reciprocal obligation systems has been supplied to support both sides of the argument (Pankhurst 1985; Rahmato 1987; Cekan 1990; Shipton 1990).

The present findings are also mixed. Roughly one-third of the households at Adele Keke, Beke Pond, Gara Godo, and Korodegaga report that they supported (or were supported by) their relatives during the famine. The highest figure, 81 percent, came from pastoral Beke Pond. At the three other sites, the share of households reporting extra assistance was only between 13 and 25 percent. The remainder said that things were so bad that they could not help anyone but themselves.

In Doma, for example, some relatives went out of their way to avoid seeing each other rather than confront the embarrassing issue of blood ties. In Dinki, where conditions were the worst of all the sites, a common response was that people felt a moral obligation to bury a neighbour if found dead, but that other forms of help were limited. One man replied, "There was no way of helping each other. It was a time of hating—even your own mother."

Although one-third of households sampled shared more food and income with relatives during the crisis than during normal times, such sharing shows up more among households with a relatively higher income. Whereas only 29 percent of the very poorest households increased their offer of, or demand for, assistance from relatives, the proportion among less poor households was 43 percent. What is more, very few households report sharing with people who were not their relatives. In other words, sharing (and, indeed, increased sharing) of resources did take place, but relatively more of the wealthier households were in a position to do so, and they shared with blood relatives rather than with destitute households outside of the family.

Where it was difficult to find access to shared resources, many households resorted to credit. Debts represent personal ties, and personal ties represent security during crisis. Thus, if relatives could not (or would not) give food or cash in the form of a gift, often they would give a loan. Forty-eight percent of loans were received from relatives, another 41 percent from friends, with only 11 percent coming from professional money lenders or merchants. There was some difference in this pattern across agro-ecological zones and income groups. For example, in the highlands an average of 58 percent of loans were to/from relatives and only 6 percent from money lenders. In the lowlands, the proportions were 47 and 14 percent, respectively, indicating both greater need and lesser support in the lowlands.

Similarly, slightly more households at the upper end of the income scale borrowed cash (37 percent) than did the poorest households (32 percent). They also made, or took, larger loans. This indicates that the poor (but, in fact, all households) lack access to the credit that is crucial to preserving resources during times of stress. Interest rates on loans ranged from 50 to 300 percent, payable in cash or kind. Where they were obtained from relatives rather than from merchants they usually carried no time limit for repayment.

Famine Impact on Food Consumption

Even during nonfamine years, food consumption levels in Ethiopia are extremely low. The figures commonly cited lie in the range of 1500 to 1750 kilocalories per person per day (Idusogie 1987; Harbeson 1990; UNEPPG 1990; World Bank 1993). It should be stressed that these are rough estimates based on fragmentary surveys, food balance sheets (which often exclude roots and tubers from the calculation), and scattered nutritional monitoring data (Mulhoff 1988; Kelly 1987).

In order to better understand the eating patterns of the sample households, questions were asked about the household's consumption of all food items during the previous week using local units of measure. The head woman of the household was interviewed wherever possible. Where this proved to be impossible a more junior woman or the male household head was interviewed.[5]

Variability existed in food consumption levels and composition both within and across sites. Table 4.6 indicates that the highest average consumption levels during 1989 were in Debre Berhan, the more prosperous of the survey sites. Indeed, only one of the three highland sites (Gara Godo) showed an average level of calorie consumption that was worse than the average for the sample as a whole. In Gara Godo, 84 percent of households were not even consuming enough food to meet the 80 percent of minimum requirement level that is commonly used as an indicator of the threshold to malnutrition (Foster 1992).

All three lowland sites, on the other hand, were at or below the sample average in recommended dietary terms. The average lowland consumption

Table 4.6. Daily calorie consumption, by survey site, income group, and agro-ecological zone 1989/90 (source: survey data, 1989/90)

Zone	Survey site	Income groups (kilocalories/person/day)			Households below 80% RDA[a] (%)	Average consumption in kg/year[b] (kg/person)
		Poorest	Middle	Less poor		
High-land	Adele Keke	1997	1905	2165	67	223
	Debre Berhan	2065	2733	3207	42	295
	Garo Godo	1366	1382	1621	84	161
Low-land	Dinki	1279	1621	2617	75	203
	Doma	1554	2384	1995	68	219
	Korodegaga	1564	1505	1845	83	181
Pas-toral	Beke Pond	2371	2147	2792	43	269
Weighted mean[c]		1693	1685	2183	68	194

[a] For this calculation the minimum recommended daily allowance (RDA) is set at 2300 kilocalories per capita per day.

[b] Daily calorie consumption per person in kilogrammes of wheat equivalents.

[c] The weighted means are calculated using income pooled across all sites, not based on the means of each site combined. ANOVA analysis of variance indicates a significant difference between the means for the income groups at three of the sites: Doma (at 5 percent significance) and Dinki and Debre Bephan (both at 1 percent).

was 1755 calories per person per day, compared with an average highland consumption of 1932 per day. This poorer showing in the lowlands conforms to their status as generally poorer communities than those in the highlands.

Taking income as the stratifier, the very poorest households were consuming 30 percent fewer calories on average than the less poor households. The gap between higher and lower ends of the income scale is greatest in Dinki, and between the groups is narrowest in Gara Godo. Yet, by any nutritional standard, all of these figures are low. Consumption levels of the poorest income group at four of the survey sites (including all three lowland sites) were below 1600 kilocalories per person per day, indicating extreme food deprivation even in the relatively good rainfall year of 1989/90. Even the less poor households in Doma, Korodegaga, and Gara Godo consumed less than 2000 kilocalories per person per day.

As a result, 68 percent of households were, on average, consuming less than 80 percent of the recommended daily allowance of 2300 kilocalories. This compares unfavourably with recent surveys in The Gambia and Rwanda, which found only 18 percent and 41 percent, respectively, of households as calorie-deficient (von Braun and Pandya-Lorch 1991). The range shown in Table 4.6 runs from a low of 42 percent in Debre Berhan to highs of 83 and 84 percent in Korodegaga and Gara Godo. In other words, in 1989/90 (a good year) no less than 68 percent of the households in these communities could be classified as malnourished.

Taking the analysis a step further, although households headed by women (widows, divorcees, or wives of soldiers) earned less than male-headed households (a net income of US$62 per person versus US$53), their expenditure on food and calorie consumption was similar. At the time of the survey, households headed by men were spending a monthly average of US$16 per person on food, whereas expenditure in households headed by women stood at US$18. Such expenditures represented roughly 66 and 67 percent of monthly income, respectively. In caloric terms, male-headed households were consuming a daily average of 1927 per person compared with 2110 consumed in households headed by women.

There was also little difference observed between the food budget shares of the poorest households and the less poor: 65 and 64 percent, respectively. These shares compare with a 59 percent expenditure on food by the poorest income groups in The Gambia in 1985/86 (von Braun, Puetz, and Webb 1989), 63 percent among landless labourers in Kenya in 1984/85 (Kennedy and Cogill 1987), and 37.5 percent for the poorest groups in Rwanda in July–September 1986 (von Braun, de Haen, and Blanken 1991). In other words, even the wealthiest households in this sample spent more on food than the poorest income groups in recent studies in other parts of Africa.

The most common sources of calories consumed by the sample households were maize and wheat, representing 51 percent and 10 percent of total calories, respectively. Sorghum and pulses were next in importance at 8 percent each. Teff, barley, and more expensive calorie sources, such as meat, oil, and sugar, each contributed less than 5 percent to the total of calories consumed.

Of course, consumption of the major staples was not identical across income groups. Higher income groups have a more varied diet than poorer households. For example, households with a higher income received more calories from barley, sorghum, and pulses (30 percent of the total) than the very poorest households (15 percent of total). Similarly, the latter ate fewer vegetables, oil, and butter than the higher income group. Conversely, the very poorest depend on maize (the most abundant and cheapest cereal) for 63 percent of their calories, compared with the less poor, who consume only 39 percent of their calories in that form.

Yet, were households with a higher income able to rely on the same foods during the crisis? There are three main consumption responses to absolute food shortage: the diet can be diversified to incorporate food items not normally consumed, the quantity of food consumed per meal can be reduced, and the number of meals per day can be reduced. All three measures were adopted by the households sampled.

The most striking difference between famine-year consumption and 1989 was a marked lowering of consumption across the sample of the most expensive grain (teff), the least expensive grain (maize), and roots and tubers. During 1989, an average of 11 percent of households were consuming some teff and 83 percent of all households ate some maize. During the famine, less than 1 percent of households ate teff and only 41 percent consumed maize. Similarly, consumption of roots and tubers was prevalent among 25 percent of all households in 1989, up from 16 percent during the famine.

For many these differences were made up by an increase in the use of wheat and the consumption of "famine foods" (Irvine 1952). Although certain forage foods are collected as a matter of course, even during normal years, the range of items and frequency of consumption at the survey sites rose considerably during the crisis. This feature of drought and famine response has been noted in many parts of Africa, such as Kenya (Neumann et al. 1989) and Niger (Kelly and Taylor-Powell 1992).[6]

In Dinki and Gara Godo, more than 95 percent of households supplemented their diets with famine foods such as roots and leaves and, in the case of Dinki, even grass and rats. At other sites the average number of households consuming famine foods was close to 35 percent. Both men and women joined in the search for products to supplement an increasingly restricted diet.

The lower half of Table 4.7 shows that 58 percent of households in the relatively higher income group increased their consumption of famine foods, compared with only 41 percent of those in the lowest group. This is because the poorest households supplement their diet with berries and fruits even in

Table 4.7. Household consumption responses during worst famine year, by sample survey site. The worst year of famine varies according to individual household responses for the period 1983–1988 (source: survey data, 1989/90)

| Zone | Survey site | Meals eaten per day (percentage of households) | | | Reduced quantities eaten | More famine foods eaten[b] |
		One or fewer	Two	Three		
Highland	Adele Keke	22	56	22	84	35
	Debre Berham	9	27	64	30	33
	Gara Godo	20	78	2	84	99
Lowland	Dinki	78	20	2	89	96
	Doma	–	92	8	31	30
	Korodegaga	12	82	6	69	16
Pastoral	Beke Pond	69	29	2	47	47

[a] Data are for peak of crisis.

[b] "Famine foods" are unusual foraged foods, such as grass, roots, and rodents.

"normal" years. Frequently consumed "collected" foods should therefore be distinguished from famine foods. The gathering of uncultivated foods is a coping mechanism, whereas the gathering of famine foods represents desperation.

RISK TAKING TO SURVIVE

The other two methods of dealing with food shortage (reduced consumption per meal and reduced number of meals) represent severe hardship and a lack of alternatives. It is here that the definitional distinction between coping and suffering becomes indistinct. When people fight over handfuls of roots, as was reported more than once in Dinki, life rather than livelihood is at stake.

Disappearing Food and People

The fourth column in Table 4.7 shows that 30 to 89 percent of households across all sites reduced the amounts of food consumed per meal during the famine. Only 30 percent of households in Debre Berhan reported such a reduction because their experience of famine was much less than at other sites. By contrast, only 31 percent of households in Gara Godo reduced the amounts consumed because they were already consuming such small amounts. Similarly, more households *reducing* the amounts eaten fall into the relatively higher income category rather than among the very poorest, because the latter were starting from such a low consumption base to begin with.

Most households also cut back on the number of meals per day. The first three columns of Table 4.7 give an indication of the extent of such reductions. The most extreme case was recorded in Dinki. In 1989, 67 percent of households were consuming at least three meals per day. During the famine, 78 percent of these same households were limited to one meal per day, or even fewer. The "or fewer" is included because more than a dozen households went up to four days without any food at all. Similar, although relatively less severe, reductions in food were experienced at other sites. In Beke Pond, for example, 69 percent of the pastoral households ate only once a day during the famine. In Debre Berhan, by contrast, only 9 percent of households were reduced to such a state of penury.

Some were able to flee the famine. As resources were used up many individuals and whole households took the chance of leaving their homes in search of assistance. Thousands of refugees crossed the border into Sudan from the northern regions, and some moved from the south into northern Kenya (Clark 1986; Kidane 1989; de Waal 1991). In 1991, an estimated 1.2 million Ethiopians were still living in Sudan, of which 300 000 were settled in camps supported by the United Nations High Commission for Refugees (IHD 1992; Mercer 1992).

The number of people relocating within Ethiopia (not counting the movement towards feeding camps) was apparently much smaller. Only 54 individuals from the sample households left their homes permanently during the crisis years of the mid-1980s, and only seven left and subsequently returned. These low figures are explained partly by the fact that each of the villages surveyed was reached by a relief organization before large numbers of people attempted to migrate. However, personal mobility was highly restricted during the 1980s (enforced by road-blocks and militias), meaning that migration was not typically a free or easy choice.

Food had returned to Ethiopia by the 1990s. Several good harvests coupled with mass shipments of food aid raised consumption levels back to their previous, minimal levels. However, many people have not returned. The journey home is as long and possibly as arduous as the journey away. Without the cash to fund a trip of hundreds of kilometres, and without the certainty of what will be found on return, many refugees sit tight.

Coping with Death

For many of those who remained in their villages, death was the last stage of the famine. Mortality statistics alone cannot do justice to the magnitude of the suffering involved. A woman in Korodegaga described 1985 in the following terms:

It was a time when there were no birds singing. They must have died too, I suppose. Vultures drifted high in the air without flapping their wings. They didn't need too since the air rising from the ground was so hot. Cows bellowed from hunger throughout the night and cockerels crowed at strange times. But, children didn't cry. That was the worst of it. Children looked their mothers in the eyes, pleading for food, but there was none. I couldn't find any and my breasts were dry. My little girl died on the fifth day without food. I should have died with her.

This woman, aged 35, lived on, but many did not. In Doma, 37 percent of sample households experienced at least one famine death. In Korodegaga, it was 45 percent, and in Dinki, almost 53 percent. Dinki's experience was the worst. Out of a total of 120 households making up that village, 18 whole families were lost. In 1990 their homes stood empty and untouched. Pots and pans lay scattered around the fire where they had been left 5 years previously.

In most cases, death came slowly, not suddenly. People conserved energy by not moving for days on end. A raid of Afar cattle-rustlers on Dinki in late 1985 met little resistance. The cattle were worth little by that time, most people had already sold their rifles in order to purchase food (for US$50 versus the nonfamine price of US$350), and men were too weak to respond anyway. Several households in Dinki went for days on end without water because no one was sufficiently strong to climb back up the steep slope from the river.

In Korodegaga, a 15-year-old boy was found dead lying under a tree where he had sat down to rest 3 days previously. This was common. In Gara Godo, it was said that: *Sabasbati chilen Gabati, duti Karati* (during the famine, people could eat a piece of pancake in the market but still die on the way home). This was a major problem in Dinki. So many people were found dead in their fields, by the river, on the path to the market, and so few people had the energy to carry the dead home, that bodies were buried where they lay. This had two repercussions. First, there was a mass conversion of Orthodox Christians to Islam in 1985. Rather than face the risk of being buried in unhallowed ground, large numbers of people became Moslems (for whom relatively less emphasis is placed on hallowed burial). This underscores the fatalism of the times.

The second repercussion was that corpses were not buried very deep in the ground. An old man, the so-called village idiot, recounted the following:

I cannot find words to explain how bad it was. It seems like a dream to me now how men and beasts disappeared. The sun burned like a piece of charcoal. There was nowhere to turn. We did not have the strength to give neighbours the Christian burial that they deserved. So I often called in people from the next village to help me dig shallow graves for people where I found them. The problem was that bodies lay barely covered by soil. I spent a lot of time chasing cows and dogs away from graves. They were gnawing on the bones of their owners. After that we decided to put people into large grain storage jars wher-ever possible and to bury those. That worked until I was too weak to dig at all. I actually pushed two people in wooden coffins into a well because no one would help me dig. I wonder if anyone would have bothered to bury me?

The most immediate cause of death in Dinki was cholera. Inhabitants of villages located, like Dinki, at the foot of the Rift Valley escarpment were told by villagers on the plateau above not to climb up the slope for fear of confinement or forced repatriation. Similarly, people in the highlands who wished to visit family in Dinki were told not to come back. In effect, a whole locality was quarantined by frightened villagers living in the relative safety afforded by physical remoteness from catastrophe.

This confirms the proposal by Watkins and Menken (1985) that, "famine seems to be associated with an increase in deaths from a number of infectious diseases" (see also de Waal 1989). In addition to cholera, there were reported outbreaks of meningitis and yellow fever in other regions. In Dinki, Doma, and Korodegaga (all lowland villages) the single greatest cause of death during the famine reported by the households themselves was disease—46 percent of a total of 136 cases. Most of these victims were children aged less than 2 years (28 percent) and adults older than 46 (25 percent).

Second in importance as a cause of famine mortality was an absence of breast milk (28 percent). Although the fact that breast milk dries up at all during famine has been challenged in the literature (Rivers 1988; Huffman 1990), women in this sample confirm that lactation can be compromised severely by prolonged reduction in calorie intake. More instances of death are ascribed to a loss of breast milk than to frank starvation.

What is more, a higher proportion of women in higher income households reported this problem than did women in the very poorest households. Across the sample as a whole, the proportion of households in which mothers were forced to wean their babies earlier than expected (because their breasts dried up) ranged from 23 percent in Adele Keke to 84 percent in Gara Godo. The average was 36 percent for all sites. But, 57 percent of women in less poor households reported the problem compared with only 29 percent among the very poorest households. That this issue was raised most by relatively higher income women may indicate that they had intended to breast-feed their infants longer than women in the poorest households. Confirmation of this hypothesis would require further research on weaning habits in normal and crisis periods.

Frank starvation was reported as the actual cause of death for the remaining 26 percent of deaths in the households sampled.

COPING WITH SURVIVAL

The aftermath of famine is not the aftermath of war (von Clausewitz 1832). No enemy has been defeated. There is no elation for the victor. Food may become less scarce again, but food insecurity remains. The next crisis looms just around the corner.

Some people certainly make gains during famine. Households able to purchase assets at low prices, and those who have a surplus of grain after the first good post-crisis harvest, did become relatively richer. But for most people there is only a marginal improvement in conditions after famine has passed, and the succeeding months represent a tough period of reconsolidation. The first goal is to eat and regain health. The second is to invest in the next harvest while seeking alternative sources of income. The third is to restock the household with productive and durable assets.

Unfortunately, some things cannot be replaced. In 1993, it was estimated that there were 90 000 orphans in Eritrea and more than 25 000 orphans concentrated in camps and church compounds in Wollo, Tigray and Shewa (Davidson 1993; de Regt 1993). These children ranged in age from young infants to teenagers, many separated from their parents on the trek between village and feeding centre during the mid-1980s. Some children left alone in the camps were reunited with relatives during the late 1980s. For example, the Ethiopian Orthodox Church had succeeded in reunifying 4300 of the children under its care with parents by 1993 (Schellinski 1986; de Regt 1993). However, thousands of orphans remain dependent on international charities and church organizations for their future.

Many children flourish in their new environment. The better camps provide schooling, vocational training, counselling, and new clothes. Older youths are helped to find jobs or to establish themselves as independent farmers on donated land. Yet, some children are physically or mentally scarred for life. Some cannot remember their name, others have not spoken a word since their ordeal almost a decade previously. Some vomit after every meal.

What is more, not all parentless children ended up in an orphanage. It was not unusual for the oldest child in a family to be left caring for younger siblings. For example, an 8-year-old girl in Dinki was faced with feeding three younger siblings after she watched her father and mother die in 1986. She succeeded in her daunting task by spinning cotton that she sold in a market 6 hours walk away. She also signed up for employment in a local food-for-work project and received 3 kilogrammes of wheat for each day of work building a road up the mountain side. That road made her trips to the market less difficult than before. She shared whatever food became available with the younger children. In 1990, all three siblings were still alive; not flourishing, but alive. Their 12-year-old caretaker was accorded the respect of an elder in the village.

NOTES

1. This was not part of the larger government initiative to move people from regions in the north to the south and west of the country. It was a spontaneous move assisted by RRC and UNICEF.
2. It also has some irrigation potential, being situated alongside the Awash river, but numerous attempts at developing such potential have failed (Webb 1989a).

3. This simple distinction between upland and lowland should not obscure the real problem of ecological zonation in Ethiopia (this is discussed in detail in Wolde-Mariam (1991)). Upland and lowland categories can be subdivided many times according to altitude, rainfall, and temperature, but there is no standard classification of zones because there is no agreement on relevant parameters.

4. Few farmers mentioned constraints relating to land titling, land fragmentation, producer prices, or government extortion through grain quotas. Although a lack of tenure rights has been widely blamed for low productivity, this may apply most to regions where security of tenure was lowest, such as parts of Wollo and Tigray (Stahl 1990; Lycett 1992). The same is true of the quota system. Less than 50 percent of the sample households paid a grain quota in 1988/89. Of those that did, 65 percent were households in the highest income group. The average quota paid by these households was only 7 kilogrammes.

5. Food consumption data relate to 400 of the 547 sample households. Data for the remainder were incomplete or unreliable. All items that were home-produced, purchased, or received as gifts (cereals, noncereals, snacks, condiments, and foraged foods), were included in the analysis. Although there was certainly some under-reporting where snacks, condiments, and foraged foods are concerned, the figures for major calorie sources are relatively reliable. It should be understood that because this was a single-shot consumption survey the time spent by households at such low intake levels cannot be reported.

6. Zinyama, Matiza, and Campbell (1990) argue that in Zimbabwe, unlike other parts of Africa, there was no increase in the use of wild foods during the drought of the mid-1980s. They attribute this to the availability of migration employment and to relatively efficient national food transfer programmes (see also Webb and Moyo 1992).

5 The Pastoral Experience

While mention was made in the last chapter of varying manifestations of famine by agro-ecological zone, the analysis focused on sedentary farm households. A separate discussion here on pastoral experiences of famine is warranted on two grounds. First, pastoralists face different constraints than farmers and may require different solutions. Africa's pastoral households derive more than 50 percent of their income from livestock products, and a substantial part of their diet from home-produced milk, meat, and blood (Monod 1975; Sandford 1983). In other words, pastoralists generate most of their economic goods and calories via the agency of livestock rather than that of crops. Remedies for pastoral food insecurity can therefore be different than for crop-dependent populations.

The second reason is that the traditional strengths of African pastoralism appear to be in decline. There is concern that drought, population growth, territorial contraction, and inappropriate development are leading to the breakdown of traditional systems (Bonfiglioli 1988; af Ornas 1990; Bascom 1990). This can lead to increased poverty among the poor, greater vulnerability to future climatic and economic crises, and accelerated resource degradation that may ultimately compromise the long-term sustainability of pastoralism as a way of life (IGADD 1990; Stone 1991).

This chapter examines the characteristics of two pastoral groups in southern Ethiopia and examines how they were affected by the drought that also affected their sedentary neighbours.

PASTORALISM AS A LIVELIHOOD

There are some 30 nomadic and semi-sedentary pastoral population groups in Ethiopia, together comprising over 3 million people (UNDP/RRC 1984; OPHCC 1991). As elsewhere in Africa, most of the pastoralists inhabit less-favoured parts of the country. These tend to be areas of low and variable rainfall (average annual rainfall between 150 and 400 millimetres, high drought frequency), infertile soils (clay and organic matter content usually less than 1 percent, limited moisture retention), and covered by sparse, specialized vegetative cover adapted to both of the above (Harrison 1987; Matlon 1987; Sollod 1990). In Ethiopia, these conditions characterize most of the semi-arid lowlands making up one-third of the country's area.

The socioeconomic systems that operate under such harsh conditions have evolved complex mechanisms for survival. These can be grouped under three headings: diversification, mobility, and communal decision-making.

First, pastoral herds usually include more than one animal species.[1] Large ruminants form the core of most herds because milk production is one of the main functions of livestock holdings. The share of milk in the diet ranges from 25 to 76 percent of total consumption according to season and ethnic group (Bernus 1980; Donaldson 1986; Grandin 1987).[2] Among the Ethiopian Borana, for example, 30 to 40 percent of the milk yield is consumed, leaving 60 percent to suckling calves (Holden, Coppock and Assefa 1992).

However, small ruminants are also kept because they have higher reproductive rates and greater disposability through sale when cash is required. Small ruminants are more disposable because their sale is usually left to the discretion of the individual, whereas the sale of large ruminants is often an event of wider importance to the community which requires consultation before action. A mixed herd also has the advantage of diversifying forage needs and of raising the average level of drought resistance of the herd (Mace 1990; Stone 1991).

Second, mixed herds are kept on the move. Herd mobility is central to both nomadic and semi-nomadic systems. Herds seek seasonal vegetative growth, capitalizing on water and grazing when and where these are available (Monod 1975; Jahnke 1982). Alternatively, milk-producing and young animals can be kept at a central point (water sources or enclosures, such as the dispersed "cattle posts" of Botswana), while older animals graze on long-range loops around that point. The Borana in Ethiopia divide their herds in this way. The *warra* herd comprises milking and young stock kept at grazing grounds close to home. By contrast, the *forra* herd, made up of non-lactating adults and immature animals, follow freer-ranging loops that bring it close to camp on 3- to 7-day cycles (Tilahun 1984; Donaldson 1986).

Third, communal decision-making is strong. Complex systems of collective land and water management have been developed to share limited resources. For example, the Borena rangelands are divided into 75 demarcated districts called *meda* (UNDP/RRC 1984). Each of these has one or more water points and a core population divided into several dozen encampments under an overall clan leadership. Although the *meda* are used communally, individual access to grazing land is commonly controlled by the clan. The Borana, like the Fulani and WoDaaBe of Niger, allocate "well keepers" responsible for arranging the time schedule and duration of watering periods allowed for each clan and household around an individual water point (Thebaud 1988). In other words, a lack of private tenure rights to resources does not imply their unbridled exploitation by individual households.

What is more, choices about herd composition, consumption and expenditure patterns, and about herd mobility generally take place within clearly prescribed physical, sociopolitical, and territorial limits (Monod 1975). Physical limits are set by the distributional patterns of resource availability, as well as by the prevalence of parasites and epizootic diseases. Sociopolitical limits are defined by rules set in clan councils, whereas territorial limits are set by the proximity of other pastoral groups and settled agriculturalists.

These key features of traditional pastoralism have lent themselves to an opportunistic utilization of resources in regions of relatively poor resource availability and low population density. Such low-input, low-output systems rely on few modern technologies, and labour (for herding and watering) is often the main factor of production.

However, competition for the limited resources available is constantly growing. Human population growth appears to be roughly 50 percent above that of livestock population growth in many areas, which stretches the gap between milk supply and demand (Anteneh 1984). Human growth also leads to the exploitation of marginal grazing areas, the encroachment of herding into traditional grazing reserves, and the expansion of cropping into more fertile grazing lands.

For example, Moris (1988) has argued that, "virtually all of Africa's major irrigation schemes have been located in what had been critical dry season grazing for pastoralists." In Ethiopia, over 65 000 hectares of land bordering the Awash and Wabi Shebele rivers were converted in the 1960s from dry season grazing into irrigated farms (Bondestam 1974; Gamaledinn 1987). Nonirrigated cultivation in marginal lands has also expanded in recent years.[3] These trends serve to further restrict the resources available during both good rainfall and drought years, resulting in higher livestock mortality at an earlier stage of a crisis.[4]

Livestock population growth also plays a role in the above, but it is more frequently accused of degrading the existing resource base. Although the trend and degree of natural resource degradation continues to be debated, it is widely argued that degradation is a serious problem in many drier regions.[5] Pastoralists have been widely blamed for overstocking, trampling of soils around water points (which reduces soil porosity and increases run-off), destruction of woodland through burning (intended to stimulate green growth), and the prevention of organic build-up in the soil by feeding crop residues to animals (OECD 1988; de Haen 1993).

Others believe that traditional systems are well adapted to their environment, economically efficient given current operating constraints and therefore not destructive until put under pressure (Monod 1975; Hogg 1992). While the potentially destructive impact of large, concentrated herds on natural resources is recognized, it is argued that the cause of concentration is population growth, which, in the absence of economic opportunity, contributes to poverty where the base-level of resources cannot be increased.

The result of growing competition is that pastoralists are forced to operate within contracting boundaries, changing long-standing migration patterns and concentrating activity around water points as demand for grazing and water rises. This increases conflict over diminishing resources, not only between pastoralists and farmers but among pastoralists themselves (Dahl and Hjort 1979). Such conflict is heightened during protracted drought and famine.

ADAPTATION TO STRESS IN SOUTHERN ETHIOPIA

Households interviewed at the Beke Pond survey site belonged to two pastoral groups. The first are the Borana, cattle herders that have been culturally and politically dominant in southern Ethiopia for much of the past five centuries (Wilding 1985). The second group is the Gabbra, ethnically related to neighbouring Somali clans, with a relatively greater focus on camel husbandry. The Gabbra were "resettled" into the heart of Borana territory by the Ethiopian authorities after the 1977 war with Somalia. Since then the two groups have coexisted relatively peaceably, sharing the same grazing and water resources.

Despite the different focus of their herding activities, the two groups have many characteristics in common. The average household in both sample populations contains roughly six persons, their consumption levels in 1989 averaged between 250 and 270 kilogrammes per person per year, and the sources and use of income were also similar. However, these similarities also put both pastoral groups in distinct contrast to the farm households considered in this study.

For example, although average net annual income per person in the sample was roughly the same for pastoralists and farmers (US$50 and US$52, respectively), pastoral income was more heavily dominated by livestock and non-agricultural sources than that of the farm households (see Table 4.4). Whereas smallholder households earned between 47 and 54 percent of their total net income from crop production, the latter generated less than 20 percent of income for the pastoralists.

Some semi-nomadic pastoralists do engage in crop cultivation. In the recent past it was only the very poorest who turned to farming, largely with a view to replacing livestock lost during drought. In the 1970s, for example, it was stated that "the Gabbra do not cultivate" (Torry 1977). By the early 1980s, however, increasing numbers of even wealthy Gabbra and Borana were cultivating maize and sometimes cowpeas on a semi-permanent basis (Gebre-Mariam and Fida 1982; Coppock 1991). Roughly 46 percent of the pastoral households in the sample had commenced crop cultivation since the 1985 famine. The remainder had already started since the 1973/74 famine.

Such cultivation has generally expanded in valley bottoms where soils are more fertile, techniques being copied from peri-urban farmers. In 1988, the average area cultivated by sample pastoralists was 0.7 hectares per household, compared with 0.5 and 0.8 hectares per household in the highlands and lowlands, respectively. The relatively greater area cultivated in the lowland and pastoral areas is due to a lower population density and poorer soils, which together encourage extensive rather than intensive cultivation.

Despite increased cropping, income from animal husbandry is still the most important for pastoralists. Livestock income represented 52 percent of pastoral income among the survey households—29 percent from animal sales and 23 percent from the sale of animal products (milk, butter, and cheese) traded mainly by women in towns within 20 kilometres of their camp (Holden, Coppock and Assefa 1992). This compared with an average of only 13 percent among farm households (9 percent in the lowlands and 18 percent in the highlands). This difference is explained largely by differences in the size of animal holdings. Whereas the pastoral households held an average of 39 tropical livestock units (TLU) per person in 1989/90, the highland farmers only had 4 TLU and the lowland farmers only 1 TLU per person at that time. This underlines the scale of the pastoral undertaking.

Yet, nonagricultural income is also crucial to the pastoral livelihood. Nonfarm income represented an average of 28 percent of total net income both for the farmers in the highlands and for pastoral households. However, the sources of such income were very different. The bulk of nonagricultural income in the highlands (some 10 percent of total household income) came from wages associated with manual labour, followed by private remittances from relatives (6 percent). The sale of fuel products contributed another 5 percent to the total net income. For pastoralists, the largest share of nonagricultural income (12 percent) came from the sale of home-produced goods such as baskets and pottery. This was followed in importance by wage employment on sedentary farms (9 percent) and wages from manual nonfarm labour (4 percent).

Although important in showing the degree of participation of pastoralists in the wage economy these figures are almost certainly too low because they underreport income from trade. Due to the proximity of the borders of Kenya and Somalia, many Gabbra households are heavily involved in contraband commerce of guns and electrical goods, and in the long-distance haulage of rock salt by camel-train. The Borana are also traders, but of domestic goods such as cloth, tea, and sugar, and on a more local scale. These activities were not reported on by the pastoral respondents who fear taxation or criminal proceedings. As a result, the overall contribution of nonfarm, non-wage income is more significant than indicated in the above data.

In 1989/90 the bulk of pastoral expenditure (54 percent) was allocated to nonfood purchases.[6] These were dominated by the purchase of clothing (46 percent), payment of taxes (22 percent), and purchase of cleaning agents (15 percent). Farm households, by comparison, allocated an average of only 34 percent of expenditure to nonfoods. Most of the latter went to the payment of taxes (38 percent), followed by clothing (25 percent) and fuelwood (14 percent).

At the same time, only 54 percent of food consumed by pastoralists sampled was purchased from the market (Table 5.1). This compares with an average of 61 percent for the farm households, ranging from 36 percent in Dinki to 95 percent in Korodegaga. The relatively low dependency among pastoral households on market foods and their greater expenditure on non-food products can be explained in two ways. First, pastoralists consumed 89 percent of their total 1988/89 crop harvest, whereas an average of only 65 percent was consumed by sedentary households. Highland farmers, for example, sold 15 percent of their harvest to derive a cash income, lowland farmers sold 11 percent, and pastoralists sold less than 6 percent (the difference representing gifts and loans).

The second explanation is that pastoralists rely more heavily on home-produced milk for their calories than do sedentary farmers. Table 5.2 shows that whereas maize is the single most important source of calories for all agro-ecological zones in the sample, milk (and milk products) comes a close second among pastoral households, who eat correspondingly fewer other cereals, but milk is a barely significant source of calories for farm households. Similar differences apply to sugar and pulses. The latter represent an average of 8 percent of calories for farm households, but are not eaten at all by pastoralists. Conversely, sugar contributes 18 percent of calories to the pastoral diet (mainly consumed in tea), but provides 1 percent or less to farm households.

The composition of the pastoral diet in 1989/90 was relatively less deficient in calories than that of the farm households. While the poorest lowland households were consuming an average of less than 1500 kilocalories per person per day, and the poorest highland farmers were a little better off at 1800 kilocalo-

Table 5.1. How calories were obtained by sample households in 1989/90, by agro-ecological zone (source: survey data, 1989/90)

Zone	Survey site	Calorie source (percentage of total calories consumed)		
		Purchased	Produced	Gift/other
Highland	Adele Keke	76	18	6
	Debre Berhan	38	62	0
	Gara Godo	72	28	0
Lowland	Dinki	36	62	2
	Doma	49	49	2
	Korodegaga	95	3	2
Pastoral	Beke Pond	54	42	4

Table 5.2. Sources of calories consumed by sample households in 1989/90, by agro-ecological zone (source: survey data, 1989/90)

| Calorie source | Agro-ecological zone[a] (percentage of total calories consumed) | | | |
	Highland	Lowland	Pastoral	Average
Cereals:				
Maize	38	62	43	51
Wheat	9	11	0	10
Other cereals	25	17	2	21
Noncereals:				
Milk/cheese/butter	2	1	32	2
Pulses	11	6	0	8
Roots/tubers	8	2	1	5
Sugar	1	0	18	1
Other	6	1	4	2
Total	100	100	100	100

[a] The highland zone includes Adele Keke, Debre Berhan, and Gara Godo. The lowland zone comprises Dinki, Doma, and Korodegaga. The pastoral zone is represented by Beke Pond.

ries per day, even the poorest pastoralists were consuming almost 2400 per person per day. As a result, only 43 percent of the pastoral households could be considered malnourished, compared with the highland average of 56 percent and the lowland average of 65 percent. Thus, the pastoralists could be seen as having survived the drought-prone 1980s in better shape than many farm households.

However, such a generalization is misleading. In 1985, almost 300 000 of famine victims registered for food aid nation-wide were pastoralists (RRC 1985a). By 1992 the national figure had reached 500 000, with 200 000 of these being fed in central Borana alone (Bocresion 1992; RRC 1992). In other words, suffering among many pastoralists was high both during the mid-1980s famine as well as in its aftermath. What is more, vulnerability to future famine increased among certain groups as a result of massive asset losses. It is this experience that we consider next.

COPING ON THE MOVE

The worst year of famine at the pastoral site was 1985. In that year, rainfall at Yavello station (closest to the survey site) dropped from a long-term average of over 600 millimetres per year to less than 350 millimetres for the third year in succession. At the end of the third year of drought, local pastoralists were facing three serious constraints.

First, their milk supply fell sharply as water and grazing became increasingly scarce. Average milk off-take per cow (local Boran species) is roughly 1 litre per day in an average rainfall year (Donaldson 1986; Holden, Coppock and Assefa, 1992). However, in 1985 milk off-take declined to an average of 400 millilitres per cow per day (Donaldson 1986). Among the sample Gabbra and Borana herds near Beke Pond, milk off-take during 1985 was reported even lower, at an average of only 150 millilitres per day (Webb, von Braun and Yohannes, 1992). It was also reported that production of camel milk remained much higher than that of cattle, averaging almost 800 millilitres per day compared with a normal average of roughly 1 litre per day. This underlines the value of camel ownership during years of drought.

Some of the declining productivity was associated with increased stresses on the cattle herds due to greater mobility and reduced watering frequency. For example, one of the responses to reduced grazing was to shift animals away from the camp on to longer grazing loops. In 1982, *warra* herds near Beke Pond accounted for an average of 71 percent of all cattle, with the remaining 29 percent allocated to the longer distance *forra* herds (Coppock 1993). However, by November 1984, 69 percent of the cattle had been moved to the *forra*.

The decline in milk availability forced a change in human diet. Compared with an average contribution of 55 percent of total calories to the diet of Borana households in 1982, the contribution of milk fell to only 15 percent in 1985 (Cossins and Upton 1987; Coppock 1993). This was compensated for by an increased consumption of blood and meat and of cereals. The Gabbra (as Moslems) generally refuse to eat the meat of animals that have died of hunger or disease. None of the surveyed households admitted to eating such meat during the famine, but 92 percent of the Borana, who are animists, did use this source of food in 1984/85. Many of the young men tending distant *forra* herds came to depend on fallen meat.

Similarly, although the Gabbra formerly drank blood as a regular part of their diet (as the Borana still do), a Gabbra council proclaimed 20 years ago that drinking blood was sacrilegious. As a result, only 25 percent of Gabbra households sampled increased their consumption of blood during the famine, compared with 42 percent of Borana households. Almost one-third of the Borana *forra* cattle were bled monthly to provide as much as 40 percent of the daily energy intake of adults (Donaldson 1986).

It is interesting that the Gabbra's taboos did not break down during the crisis. That pastoral households eschew certain foods even during a famine was noted among the Afar and Arssi of northern Ethiopia, who refused to eat fish during the 1973/74 famine (Kloos 1982). This stricture was not the case with followers of the Ethiopian Orthodox Church, such as most households in Debre Berhan, Dinki, and Gara Godo. Orthodox fasting rules, which dictate that meat and dairy products be avoided on 150 days of the year, were suspended by all households during the crisis (Dirks 1980; Selinus 1971). They have since been readopted.

In addition to blood and meat, the largest substitute for milk during the famine was grain. Cereals provided an average of 32 percent of total calories in 1982 among Boran households, a proportion that rose to an average of 52 percent in 1985 (Donaldson 1986; Coppock 1993). Little of this was home produced. Pastoral crops failed during the third year of drought, just as they did in other parts of the country, leaving stores empty of home-produced grain. Thus, most cereals came as food aid (10 percent of sample households received free food in the form of wheat flour), or more especially as market purchases. Unfortunately, the latter were severely affected by a rapidly declining purchasing power.

The purchasing power of households (their ability to acquire food) is determined largely by the terms of trade between food and other marketed products, such as livestock, cash crops, firewood, and labour. During crises the terms shift in favour of households able to sell, rather than purchase, cereals. For example, the value of nonstaples declines during drought because demand falls along with income. Similarly, the value of wage labour falls because the total area cultivated contracts during drought and demand for labour is reduced. Perhaps the most striking example of shifting purchasing power is the erosion of livestock prices against grain prices.

Because of seasonal fluctuations in animal milk supply, pastoralists supplement their diet with purchased grain even in normal years. They do this by converting high-value meat energy into cheaper grain energy at a beneficial exchange rate for the animal owner (Donaldson 1986; Cossins and Upton 1987), but drought changes this balance. As resources diminish, households reduce herd size by selling small ruminants and nonessential cattle, such as older males and young animals unlikely to survive stress. Many of these animals are destined for live export. The volume of sheep exported from Ethiopia, for example, rose sharply during the drought of the mid-1980s from less than 1000 tons in 1982/83 to more than 4000 tons during the famine year of 1984/85; the total declined again to 1900 tons in 1987 (National Bank of Ethiopia 1988).

However, mass sales saturate the market leading to a decline in the value of livestock at a time when cereal prices have begun to rise in response to increased demand. For example, the value of live sheep in Dessie (Wollo province) compared with the market price for maize fell by almost 75 percent from December 1982 to December 1984. The ratio returned to pre-famine levels by the end of 1986, only to climb again to more than twice its 1982 value during 1987/88 when another drought caused cereal prices to rise.

In such circumstances the eroded livestock–cereal terms of trade puts herders at a disadvantage in the market when their dependence on purchased cereals is increasing. At Beke Pond, 90 percent of the pastoral households sold livestock. The 10 percent that did not were among the most wealthy or the already destitute. Where possible it was small ruminants and older, less productive cattle that were disposed of. However, as conditions worsened calves and cows were also sold or slaughtered. By the end of 1985, 60 percent of the households sampled had sold or slaughtered calves, 25 percent had disposed of cows, and 20 percent disposed of bulls or heifers. According to Donaldson's (1986) survey, sales in five Borana encampments represented roughly a 20 percent reduction in total herd size.

Where at all possible, the seller tried to find a buyer within the clan so that animals rarely left the greater fold. Nevertheless, large numbers of animals did end up on the open market. This caused the real value of livestock to fall from 1 birr per kilogramme in 1982 to 0.3 birr per kilogramme in 1984; the maize price rose from 0.4 birr per kilogramme to over 1 birr per kilogramme over the same period. This represented a decline in pastoral purchasing power of almost 90 percent (Coppock 1991).

Of course, if livestock owners do not sell in time, even at low prices, they may face losing everything if the animals die. This potential loss of capital was the third major constraint facing pastoralists. Livestock mortality was considerable in many parts of the country. In Eritrea it is estimated that up to 70 percent of grazing herds were lost during the drought (Cliffe 1989). In Wollo, cattle numbers fell from 2.7 million in 1980/81 to only 1.3 million in 1985/86, and camel numbers declined from 20 000 to only 2000 during the same period. These figures are considerably higher than the 1 million cattle loss reported for the whole country during Zimbabwe's catastrophic once-in-a-century drought of 1991/92 (Nyoni 1993).

In Sidamo, Donaldson (1986) estimated a net reduction in herd size in five Borana encampments of 30 percent from 1983 to 1985. This level of reduction is confirmed among our Beke Pond sample households. Table 5.3 shows that the pastoral households, as well as the lowland farm households, lost an average of roughly 35 percent of their 1985 herd through animal deaths. "Less poor" households lost more in absolute and relative terms than the very poorest pastoral households because the former had much larger herds to begin with, and came out of the famine period with many more animals still alive. The same pattern applied to animal deaths during 1988/89, but mortality relative to prevailing herd size was much lower.

Thus, if pastoralists lost approximately 35 percent of their herd to death, and another 20 percent to sales and slaughter, total herd inventory would have been reduced by some 55 percent. This compares well with the 60 percent herd reduction estimated by Coppock (1993). It takes an estimated 3 years for a herd to recover from a 20 percent decline in total stock numbers; it takes 30 years to rebuild a herd that has been reduced by 60 percent (Toulmin 1985). In other words, losses of income and assets among pastoralists were on a huge scale.

Table 5.3. Livestock mortality among sample pastoral households during 1985, by income group (source: survey data, 1989/90)

Zone	Income group	Livestock mortality (percentage of herd)	
		1984/85	1988/89
Highlands[a]	Poorest	55	n/a
	Less poor	84	n/a
Lowlands	Poorest	30	2
	Less poor	41	8
Pastoral	Poorest	31	3
	Less poor	40	11

n/a, data not available.

[a] This calculation excludes the Debre Berhan survey site for which the data were incomplete.

The most immediate human impact was a decline in food consumption. After Dinki (the worst-affected survey site), Beke Pond reported the highest share of households (69 percent) forced to cut consumption down to one meal or less per day in 1985 (see Table 4.7). At the same time almost half of the pastoralists reduced the quantity of food eaten in that one meal, and half increased their consumption of famine foods.

Considerable sharing of resources among households took place during the worst periods. Pastoral communities are widely regarded as needing to be more mutually supportive than their agrarian counterparts (Laughlin 1974; Dahl and Hjort 1979; White 1984; Robinson 1989). For example, although access to water and grazing became restricted during the drought, 90 percent of the pastoral respondents reported that there was little friction over access to such resources. The few who did report clashes over grazing noted that this did not occur among the Borana or Gabbra, but with other ethnic groups such as the Gujji, who attempted to move into the Beke Pond area during the drought. Those reporting friction over water said that this occurred mostly among the Borana because of strict rationing imposed by the elders on access to ponds and traditional wells. Most disputes were over the allocation of water rights by clan.

Within clans, however, mutual support was strong. Over 80 percent of households supported anyone in their clan who needed assistance, as well as relatives outside of the immediate encampment. For example, 82 percent of food loans were arranged between relatives and clan members, with only 18 percent being based on commercial terms with external agents. This close-knit behaviour was recorded by Robinson (1989), who notes that, "the Gabbra remember that during the 1890s, they had assisted the Borana, who had lost most of their cattle to rinderpest, with gifts of camels. Reports . . . indicate that the Borana once again made official pleas in 1984 to the Gabbra for camels and that a number of Gabbra herd owners responded affirmatively to these requests" (Robinson 1989).

This strong community network played an important role in minimizing human mortality and emigration, both of which were insignificant at the survey site. However, the loss of assets and income had a long-lasting effect on the area. The relatively minor droughts of 1990 and 1991 pushed several hundred thousand pastoralists to seek food aid from NGOs and government offices. Many more migrated south into Kenya to seek improved grazing resources. Thus, poor pastoralists, exposed to herd loss through drought, productivity loss through reduced access to resources, and income loss through large fluctuations in the livestock–grain terms of trade, are probably more vulnerable to famine in the 1990s than they were in the early 1980s (IGADD 1990; Stone 1991; Hogg 1992).

A PASTORAL FUTURE?

Pastoralists are supremely adaptable; they have been coping with pressures on their social and economic systems for centuries. But, traditional coping mechanisms are increasingly strained by evergrowing competition for limited resources. This raises the question (discussed in more detail in the next chapter) of what role policy or project interventions might play in reducing such risk. The track record of livestock development initiatives in semi-arid Africa is not good. There is a consensus that the majority of initiatives between 1960 and 1990 met with limited success (Swift and Maliki 1984; Hogg 1985). Some initiatives are blamed with causing more problems than they solved (Ellis and Swift 1988).

For example, many policies and projects have been based on the assumption that a significant improvement in livestock and rangeland productivity per capita could be brought about if appropriate technical innovations were introduced to, and adopted by, pastoralists. Vaccination and veterinary campaigns have achieved significant control of major epizootic diseases (such as rinderpest, black-leg, and anthrax) and livestock pests (ticks) (Swift 1988). Construction of boreholes, ponds, and wells has been one of the most simple and effective interventions to safeguard and expand access to forage supplies (Bille and Assefa 1983; Sandford 1983). Forage supplements, such as crop residues, grass hay, and cultivated legumes have shown potential for maintaining and even improving calf nutrition (Coppock 1991; Coppock and Reed 1992).

However, despite proven technical feasibility, few innovations have resulted in large-scale improvements in pastoral productivity. Most vaccination and parasite reduction activities have proven to be unsustainable because key products are imported and delivery systems have tended to function well only intermittently. To be effective, such activities need to be supported by extensive delivery, extension, and monitoring services, all of which are constrained by central government funding, limited institutional coordination, and lack of training (UNECA 1985; Brokken and Williams 1990). In Ethiopia, for example, acaricide for dipping against ticks and drugs for vaccination against foot-and-mouth were unavailable for much of the 1980s and early 1990s because of foreign exchange constraints and the limited coverage of veterinary services.

It can be argued that effective "drought-proofing" of the pastoral economy has not occurred because most innovations aim to raise productivity rather than to reduce risk. Achievement of the latter requires a rethinking of objectives. Semi-nomadic agro-pastoralism is likely to remain the primary economic activity of southern Ethiopia for the foreseeable future. This holds true for most parts of semi-arid Africa. No simple, high-impact technologies exist that can make pastoral environments more productive in the short term. This is a reality that should guide policymaking for the next several years.

The question is how to achieve food security in such apparently unconducive conditions? If it is accepted that some form of pastoralism should be supported in order to maintain people in semi-arid regions, new policies and programmes must focus on reducing the risk of output losses, even at current low levels of productivity, rather than on increasing productivity and the commercial value per unit of animal. Projects that increase the live-weight of calves or fatten animals to maximize meat production or sales are not necessarily a priority for herders whose food security strategies generally preclude animal sales except in times of need. By contrast, pastoralists commonly accord higher priority to milk production, calf survival, enhanced endurance of mature stock, and improved terms of trade between livestock and grain in their search for household food security.

A new focus is therefore needed on increasing inputs to pastoralism rather than on outputs. Traditional low-input, low-output pastoral systems are relatively well adapted to the marginal environments in which they function. However, such systems are only well adapted if their mobility and access to resources are unhindered. This is rarely the case. Thus, the survival of pastoralism, and therefore to food security in semi-arid lands, rests on a reorientation of development investment on human productivity rather than on animal or biomass productivity. Despite decades of failed projects and programmes, investment in pastoralism should continue because it represents the only viable short-term economic security for most inhabitants of drier areas. However, economic growth objectives should be subservient to human food security objectives, and indicators of food security should become the real measure of successful development.

NOTES

1. It should be remembered that Ethiopia has the highest ratio of livestock per capita in all of Sub-Saharan Africa. It alone accounts for 19 percent of all cattle, 16 percent of small ruminants, 8 percent of camels, and 57 percent of equines on the continent (ILCA 1991).
2. Beyond milk and meat production, livestock are used for traction, fibre and manure production, transport, investment and insurance, and as a major source of income.
3. In Niger, total area under cultivation increased by more than one million hectares between 1976 and 1986, with 40 percent of the expansion made at the expense of fallow land and pastoral grazing grounds (Niger 1991).
4. Some grazing lands are being physically enclosed to prevent these forms of encroachment. While this is a localized phenomenon, there are signs that in Zimbabwe, Ethiopia and Mali small plots, often linked to water sources, are being fenced through indigenous initiatives to protect reserve grazing (Toulmin 1983; Thebaud 1988).
5. Resource degradation involves a decline in biological productivity of a tract of land as characterized by decreased herbaceous cover, increased occurrence of undesired forage species, bush encroachment, soil erosion, depletion of soil nutrients, soil compaction, or falling water-table (Glantz 1977; OECD 1988; Mortimore 1988).
6. This figure lies between the 70 percent expenditure on nonfood items reported among the Borana in a good rainfall year, and 34 percent reported for a drought year (Donaldson 1986; Cossins and Upton 1987; Coppock 1993).

6   Attempts to Reach the Poor

Famines are costly. The private cost to households that fail to cope with extreme stress was discussed in the last two chapters. But costs are also high for national exchequers and the international community when they decide to intervene on behalf of the poor. This chapter therefore turns to the public response to famine, its impact and its costs.

PUBLIC RESPONSE TO FAMINE

Recognition of the multifaceted nature of famine causation (and prevention) has given impetus to varied public attempts at protecting food entitlements through transfers, not only of food but also of income through employment and assets. It has also confirmed that sustainable public action depends on sustainable economic growth that addresses the poverty at the root of famine.

However, even the best-designed and managed feeding camps and food distributions are tragic manifestations of failure; the failure of governments, nongovernmental organizations (NGOs), and international agencies to have acted more appropriately in the past. Public agencies around the world are awakening late to the realization that *not* preparing against crises can be more expensive in the long term than responding to individual crises when they occur.

For example, famine relief expenditure in India during crisis years of the 1960s represented between 40 and 100 percent of the annual development budget of states such as Bihar, Gujarat, and Rajasthan (Torry 1984). Famine relief spending for Mali by the United States Agency for International Development reached US$46 million in 1984/85 compared with that donor's annual investment budget for Mali of US$15 million (Devres 1986). More costly still was the international response to the 1991/92 drought in southern Africa, which exceeded US$730 million (Interaction 1993). In other words, a lack of long-term donor investment in famine prevention can lead to substantial short-term expenditure on mitigation.

The same applies to national investment in famine relief versus prevention. When famines occur, governments are often blamed for a lack of prior commitment to building the capacity for crisis prediction and timely intervention (Gill 1986). Some of this "blame" is often coloured by political bias, but the point is a substantive one (Penrose 1987; Fraser 1988). Without adequate investment in information gathering and analysis, and in the mobilization of resources for rapid response, famine cannot be dealt with adequately. However, it has been argued that "disaster preparedness is probably one-third the province of the international community and two-thirds the province of the Ethiopian Government" (Hindle 1988).

The Ethiopian Government's task has been a heavy one. It has faced at least four major food crises and numerous minor crises since the mid-1970s. At least 2 million people have been in need of assistance each year of the 1980s. Given the country's structural food deficits, its lack of foreign exchange, poor infrastructure, high defense spending, and absence of a clear disaster-response strategy, famine has been a major problem for the government to deal with (Asfaw 1988; Lancaster 1990).

In 1980/81, for example, the government's budget for relief expenditure stood at 9 million birr (US$4.5 million). In 1984/85, this reached almost 65 million, dropping to 22 million birr during 1991/92 (National Bank of Ethiopia 1989; World Bank 1993). Domestic funds were available. What is more, there was no lack of warning about the 1984/85 famine. The role of the government's Relief and Rehabilitation Commission (RRC) has been widely commended (Cutler and Stephenson 1984; Jansson, Harris and Penrose 1987; Goyder and Goyder 1988).

What was lacking was the conceptual, financial, and institutional preparedness required for effective response to the warnings. Ethiopia's experience has tragically underlined the necessity of strong administrative linkage between institutions that identify a problem and those whose responsibility it should be to prevent and, if necessary, remedy that problem. The pre-positioning of response measures, combined with monitoring of preventative measures, is essential to famine mitigation. The alternative is begging on an international scale.

The government's inability to cope with famine led repeatedly to requests for external assistance. The international community has therefore played a key role in the public response to famine. The key players in this effort operate at two levels: through multilateral agencies and NGOs.

All major international relief organizations are represented in Ethiopia. Between 1985 and 1990, they provided almost 3.5 million tons of emergency food to Ethiopia, accompanied by more than US$250 million of nonfood assistance (UNEPPG 1990). On a much smaller scale of operation, but perhaps of greater visibility in the field, were the activities of NGOs. In May 1985, there were 48 NGOs operating relief projects under the auspices of the RRC. These engaged in activities ranging from relief measures to developmental programmes. During the famine the most common operations were medical operations and feeding programmes (intensive or supplementary feeding, and dry ration distribution). From 1985 to 1989, the NGOs (working with Red Cross agencies) were responsible for distributing between 60 and 75 percent of all food aid sent by the international donors (UNEPPG 1989b,d).

Second in importance were logistics operations (aid handling, storage facilities, transportation). Long-term rehabilitation programmes necessarily had a low priority during the crisis period. Nevertheless, these short-term activities were costly. The major NGOs (some 23 agencies) spent over US$130 million in-country between 1984 and 1988 (RRC 1990b).

By 1987, when the peak of the crisis had passed in most regions, NGOs were faced with the question of where to redirect their efforts. The answer for most was to move from relief to rehabilitation. This raised questions about long-term dependency relationships between beneficiary populations and local NGOs (Curtis, Hubbard, and Shepherd 1988; Elizabeth 1988). However, continued NGO presence in the field was crucial to the successful mitigation of the 1988 crisis. In that year, major NGOs and church organizations joined to form the Joint Relief Partnership. This organization supplied food to over 1 million people during 1988/89 (UNEPPG 1989a).

In the mid-1990s, there were still more than 50 NGOs operating, of which more than half implemented projects in collaboration with the RRC. The total number of NGO/RRC projects had fallen since 1985, and the focus of such activities shifted from relief efforts to longer term development projects. However, the fewer, more development-oriented projects still represent large capital investments: project costs for activities started since 1986/87 and ongoing in the first half of 1990 exceeded US$100 million (RRC 1990b). Total beneficiaries were estimated at more than 3.5 million.

INSTRUMENTS OF RELIEF AND REHABILITATION

A wide range of projects appeared in the 1980s, most aimed either at securing nutritional improvements among vulnerable groups or at protecting their resources. However, the means and time-frames adopted by various projects were often quite different. For example, the goal of emergency feeding programmes has the short-term function of preventing people from dying. By contrast, the goal of programmes introducing (or replacing) agricultural assets is to regenerate local production, thereby reducing vulnerability to famine. Developmental activities, such as soil and water conservation, focus on long-term natural-resource preservation and enhancement.

None of these interventions is mutually exclusive; like the components of a household coping strategy, the many parts of a public strategy are integral to overall security objectives (RRC 1991). According to the Ethiopian Red Cross Society, relief agencies choose the type of intervention according to the "period of deprivation, . . . population movement, . . . the degree and level of health and nutritional deterioration, the expected or actual food deficit and relevant information gathered during community surveys" (Yitbarek 1988).

Knowledge of the latter conditions represents an ideal case. In practice, there is no time to collect all of these data and to assess their message. Famine is characterized by urgency at all levels, from the household to the international community. Thus, most operations attempt to influence the calorie consumption of target individuals. At other times, when the emphasis is on rehabilitation, priority is placed on actions that replace or enhance productive capacity.

But which interventions succeed best in targeting vulnerable groups? And, which do so in optimal economic and human terms? Too few agencies ask both questions. A vital principle of relief activities is to maximize the relief provided among those who need it most. The RRC concurs with this view, arguing that "the primordial task in relief operations is the selection of beneficiaries. Proper selection makes relief work effective. Improper selection makes the work not only ineffective, but meaningless as well." (RRC 1985b)

The following sections concentrate on the targeting and impact experience of relief and rehabilitation activities at the seven survey sites. This is not a conventional economic cost-benefit evaluation. Rather, the focus here is on how projects were organized and implemented, which households participated and why, how well projects supported their private coping initiatives, and how things might have been done differently.

REACHING VULNERABLE GROUPS

The projects examined at the survey sites fall into four broad categories with six variants, as follows:

- Direct food transfers   Feeding camps
                              Food aid distribution
- Public works   Food-for-work
                              Cash-for-food
- Asset transfers   Ox–seed distribution
- Technology transfers   Single–ox plough

The following analysis considers each intervention type in turn, starting with an examination of how activities were organized in the field and the impact on households participating in such activities.[1]

Feeding Camps

The principal goal of relief is to save lives. Thus, the most urgent famine intervention (in a situation of late public response) takes the form of emergency feeding camps. At the height of the 1984/85 famine almost 300 such camps were distributing dry and wet food rations to the seriously malnourished. Camps were opened at two of the survey sites; namely Doma (in the lowlands) and Gara Godo (in the highlands). These were not large camps on the scale of Korem, where in early 1985 over 10 000 children were fed daily

(Appleton 1988). Nevertheless, they each attempted to rehabilitate thousands of malnourished children and mothers.

The Doma camp was opened in 1986 by UNICEF in response to deteriorating nutritional conditions. Malnourished children were selected for wet feeding according to weight-for-height. Pregnant women, the sick, and the elderly were also fed. Mothers were permitted to sleep with their children in the camp, and it was they who prepared the food (oats and canned sardines). Unfortunately, there were problems of palatability, because neither food was common to the area. This was partially overcome by mixing the fish into sauces, and mixing the oats into a sorghum and maize batter for pancakes. Only 16 children died during the 4 months of operation, out of more than 200 children admitted.

The scale of the problem was greater in Gara Godo. Opened in May 1985 by Redd Barna, this camp admitted only children less than 70 percent of weight-for-height. These children were given intensive medical care until they exceeded the 70 percent mark again. At that point, they were released from intensive care but were monitored during a period of supplementary feeding, followed by a monthly check-up during which a family food ration was distributed.

For 1985 as a whole, an average of 105 marasmic children were under critical care and intensive feeding each month. More than 70 percent of those admitted to critical care were less than 65 percent of weight-for-height (Jareg 1987). Another 209 relatively less malnourished children benefited from supplementary feeding, and over 200 family rations (60 to 75 kilogrammes of grain, plus 3 kilogrammes of supplementary food for each child under five) were given each month. The principal foods provided were wheat flour and soy-fortified sorghum and fish powder. Palatability was a problem with the fish powder and sorghum, which were later given only to children less than 9 months old (Jareg 1987).

The brunt of the crisis passed in the second half of 1986. By June 1986, the proportion of children below 80 percent of weight-for-height had fallen to less than 3 percent from a high of 35 percent in July 1985 (Redd Barna Records). Relief activities were suspended in September 1986, and attention was turned to projects aimed at the rehabilitation of famine-affected households.

Unfortunately, the improvements were not sustainable. The return of drought in 1987 and 1988, coupled with malaria and meningitis epidemics, raised the rate of severe malnutrition back to 18 percent by mid-1989. As a result, emergency feeding activities were resumed at the end of 1988. Some 10 000 packets of oral rehydration solution were dispensed during 1988, with over 1000 children being supported by food rations amounting to 3000 tons of grain and 100 000 litres of oil (Redd Barna 1989).

During 1985, at least one child was admitted to the camp from 35 percent of the survey households. Two-thirds of these households came from the middle and lower income groups. Yet, 32 percent came from households in the top income. This tends to support the argument that variables other than wealth, such as hygiene and disease, play a major role in determining nutritional status and famine impact (de Waal 1988; Kennedy and Cogill 1987).

Most children admitted remained under intensive care for 6 to 9 months. Only one child from the households sampled remained in the camp for more than 1 year, returning home only when the camp itself was closed down. Only two children from the households sampled died in the camp.

Most parents were satisfied that the intervention had saved lives. Almost 50 percent of respondents in Gara Godo remarked that "all would be dead" had it not been for the camp. The other 50 percent felt that without the intervention more people would have died, but that there were problems. For example, too many patients contracted infections while in care. Over 56 percent of respondents referred to the danger of contagion resulting from overcrowding in the camp. Although cholera and other epidemics were avoided through mass vaccinations (2000 children were vaccinated by October 1985), half of those admitted contracted some disease during their stay (Jareg 1987; Appleton 1988). Most were of an enteric nature, probably due to poor sanitation facilities. During 1985, water and firewood were collected for the camp by volunteers in the community. It was not until late 1986 that public latrines were dug, and none of the springs were protected from contamination (Jareg 1987).

The second complaint concerned communications between project and community. Although the selection process followed standard anthropometric criteria, not all project staff had the time (or local language fluency) to explain routines and rationales to the community at large. As a result, the role of mediator between project and community fell by default to the perimeter guard. Given the frustration and anxiety of the community, the guard standing between the admitted patients and the excluded became a focal point for anger. One report notes that the guard was "continually in danger of being attacked himself, and the field staff reported difficulties in bringing under control large numbers of desperate people during distribution of family rations" (Jareg 1987).

The third problem related to the family rations. Many households felt that such rations were too small. Households larger than five people received 75 kilogrammes of cereals per month. This ration was based on an assumed average household size of six people. However, the average household size in the present survey sample was almost eight people. Larger families therefore found it difficult to cope.

These problems have been acknowledged by the NGO that operated the camp, and real efforts have been made to analyze the 1985/86 experience in order to establish contingency plans for the future (Jareg 1987; Redd Barna 1989).

Free Food Distribution

Despite the unresolved debate concerning the cost effectiveness and morality of food aid, few voices were raised in 1984/85 against a mass mobilization of emergency food to Ethiopia.[2] The unprecedented scale of the public and private charitable response is well known. The high profile activities of organizations such as Band Aid were just the tip of the iceberg. For a short time Ethiopia became the Biafra of the 1980s, just as Somalia became the Ethiopia of the early 1990s.

Households at all the survey sites received food aid during the famine years, but at different scales according to need. In Dinki, for example, 95 percent of respondents received some aid, compared with less than 2 percent in Debre Berhan. The frequency and volume of food received also varied by site. Debre Berhan received food only once, during 1986. By contrast, Beke Pond, Dinki, Gara Godo, and Korodegaga (all lowland communities), received at least some food aid during four successive years. In Korodegaga, a majority of households received food aid during the 1981/82 crisis, and some continued to have access to food aid throughout most of the decade.

Distribution procedures varied. In some cases anthropometric measures were used to guide distribution (Young 1986; Shoham and Borton 1989). In others, food went to people who were "skinny in their appearance, and that had nothing more to sell" (Erni 1989). In some instances, lists of the needy were drawn up by village leaders, and it fell to them to ensure an equitable distribution of the food (Intertect 1986). There is little evidence from the survey sites that such distribution was manipulated in favour of wealthier households. Households that cut their consumption back to one or two meals per day during the crisis received on average twice as much food as households still consuming three meals per day (Table 6.1).

Table 6.1. Food aid received by survey sites during worst famine year, by meals consumed per day. The worst year of famine varies according to individual household responses for the period 1983–1988 (source: International Food Policy Research Institute survey data, 1989/90)

		Number of meals consumed per day during famine (kilogrammes of aid/household/year)		
Zone	Survey site	One or fewer	Two	Three
Highlands[a]	Adele Keke	48	85	175
	Gara Godo	223	207	50
Lowlands	Dinki	140	221	100
	Doma	92	89	33
	Korodegaga	173	159	100
Pastoral	Beke Pond	378	333	20

[a] Debre Berhan is not included because only one household there received food aid, on a single occasion.

On the other hand, the amount of food aid received was small. Standard rations recommended by the RRC were 15 kilogrammes of cereals and 600 grammes of oil per person per month, equivalent to roughly 1700 kilocalories per day, still 25 percent below the recommended minimum intake of 2300 kilocalories (Yitbarek 1988). This "ideal" ration was rarely provided for more than a few months. Households received an average of only 180 kilogrammes in the worst year. Only at three sites did households receive more than 200 kilogrammes for the year, and this in itself is insufficient to sustain a family of six for more than 60 days.

In other words, food relief in the survey areas appears to have been designed to protect household consumption for a limited time only. It could not on its own support households through the entire crisis. Instead, food aid acted as a temporary measure until other, more far-reaching interventions could be set in place. In Dinki, for example, food aid was distributed for 6 months in 1985, prior to the implementation of a more complex asset-distribution project. When the latter was established, food aid stopped. The implications of a lack of overlap between such interventions is taken up later.

EMPLOYMENT SCHEMES

Ethiopia has a wealth of experience with employment programmes, also called public works. For example, during the 1980s and early 1990s it hosted the largest food-for-work programme in Africa. It also served as a testing ground for the first large cash-for-work programme in Africa (operated by UNICEF). These programmes use different approaches to the same problem: namely, how to secure employment and adequate consumption levels for participants, while at the same time generating sustainable assets. The food-for-work projects implemented at four of the survey sites offered food as their wage, whereas households participating in cash-for-food at three sites received monthly cash wages.

Food-for-Work

Food-for-work (FFW) activities were examined at four of the survey sites. The largest was administered by the World Food Programme (WFP) in collaboration with the Ministry of Agriculture and RRC. From 1980 to 1990, WFP contributed over US$230 million to labour-intensive projects in food-deficit areas (Holt 1983; UNEPPG 1987, 1989c; WFP 1990, 1991b). Its most recent phase (US$78 million) was aimed at rehabilitating 2.6 million hectares of degraded land by offering 27 million workdays per year (WFP 1989).

The Adele Keke site examined here is in the Alemaya catchment of Hararghe Province. This catchment had the highest level of food utilization in Hararghe during the late 1980s, providing work for almost 1500 people during

9 months of the year. Payments totalled 2500 tons of grain and 99 000 litres of oil in 1988 (Vallee 1989; WFP 1991b). The project, initiated in 1984, involved soil bunding, contour terracing and afforestation.

The other three projects considered were implemented by NGOs. In 1986, NGOs ran about 60 food-for-work projects nationwide, most of which were emergency relief activities; only 4 percent of these projects were initiated prior to 1985 (Hareide 1986). By 1989, the number of projects had risen to 75, but most had been transformed into seasonal employment projects, regular development activities that focused on road-building and/or resource conservation (WFP 1989).

One of the emergency projects was initiated in Dinki. Started in June 1985, its initial goal was to provide food to famine victims while also improving local road communications. The first car to drive to Dinki arrived in June 1986, heralding improved access to food and medical care. After October 1986, the focus turned to afforestation and bunding. The project ended in June 1988.

A second project considered was implemented in Gara Godo. Started in 1988, it covered 6000 households in 30 communities (Redd Barna 1989). During 1988, 600 men and 200 women provided 23 000 working days for road building, tree planting, and gully bunding; their remuneration totalled 68 tons of maize (Redd Barna records).

The third NGO project examined was implemented in Sidamo by CARE with the International Livestock Centre for Africa and the Ministry of Agriculture's Fourth Livestock Development Programme. This programme was unusual because the participants were pastoralists and because its activities were often innovative. Food was offered for digging ponds, maintaining wells, building cement-lined water tanks, and hay-making. The latter was new to pastoralists, who therefore learned the benefits of hay-making and storage in advance of the long dry season, while at the same time earning food for their labour.

Participation in Food-for-Work

The employment provided by food-for-work (FFW) projects differed by site according to type of activity, criteria for participant selection, and local response to the food wage offered. Table 6.2 shows the days spent by households on FFW activities between 1985 and 1989. The average annual total across the four sites was 83 days. This compares closely with a figure of 90 days estimated for Ethiopia by Holt (1983) and the average of 86 days observed by Kohlin (1987) in Wollo for 1986. However, FFW was more important in Dinki and Adele Keke than at the other sites. Households in Dinki provided 150 days to FFW during 1985, indicating the extreme nature of the crisis and the lack of alternatives for households seeking food. In Adele Keke, the participation peak was in 1986.

Table 6.2. Average working days (average days/year/household) spent by household members on food-for-work activities, by survey site, 1985–1989 (source: survey data, 1989/90)

Zone	Survey site	Main project activities[a]	1985	1986	1987	1988	1989
Highland	Adele Keke	Bunds, roads	59	112	58	104	108
	Gara Godo	Roads, trees	30	95	65	60	60
Lowland	Dinki	Roads, bunds	150	112	109	141	–
Pastoral	Beke Pond	Hay-making, ponds	90	83	60	60	78

[a] Bunds, bunding (with soil or stones) of gullies and steep, degraded slopes; roads, rural road construction and maintenance; hay, cutting and drying of hay for long-term storage; ponds, excavation of ponds; trees, nursery seeding and transplanting of grown seedlings.

Many projects accept that all households wishing to participate should be permitted to (EEC 1989). This principle assumes that FFW is a self-targeting intervention from which the wealthy voluntarily exclude themselves (Traore 1989). Given the generally low levels of income in the study population, targeting may not be a particularly relevant issue in the context of the study sites. Nevertheless, it is interesting to consider the access of the poorest to FFW. Few projects made an attempt to target the poorest. A survey by Kohlin (1987) of projects in Wollo found that only 30 percent of 228 respondents felt that the poor had priority access to FFW. Admassie and Gebre (1985) similarly found only 17 percent of respondents reporting a recruitment bias toward the poor.

If that is the case, who does participate, and why? Admassie and Gebre (1985) found no uniform system of recruitment in the 24 projects in Ethiopia. The same was true for the present survey. On the one hand, at no site were female-headed households given priority. On the other hand, in Dinki and Beke Pond all households were accepted, even if the only family member strong enough to move stones was a young girl, whereas in Gara Godo only households with able-bodied members were registered.

In Adele Keke, although the Ministry of Agriculture/WFP officially accepted all households wishing to work, recruitment was managed by village leaders and a ministry foreman, based again on "ability to work" coupled with undefined criteria set by the foreman. Seventy-two percent of the nonparticipants had applied but had been rejected according to the latter "criteria". All 72 percent claimed that registration was unfair, referring to bribery, kinship, and personal connections as the real criteria guiding selection.

Although allegations of corruption always surface in connection with FFW the present analysis shows that at three of the four sites, participation by the poorest was not significantly lower than in less poor households, even taking household dependency ratios into account (Table 6.3). In general, a picture of

Table 6.3. Household participation in food-for-work at selected survey sites, by income group (source: survey data, 1989/90)

| Zone | Survey site | Income groups (percentage of participants)[a] | | |
		Poorest	Middle	Less poor
Highland	Adele Keke	24 (54)	40 (60)	36 (55)
	Gara Godo	31 (57)	39 (54)	36 (55)
Lowland	Dinki	34 (50)	34 (48)	32 (32)
Pastoral	Beke Pond	33 (50)	27 (60)	40 (55)

[a] Dependency ratios (the percentage below 15 and above 60 years of age in the total population) are in parentheses.

rather equal access emerges, with the exception of Adele Keke. In Adele Keke (where complaints of unfair selection were the strongest), only 24 percent of participants were poorest households, compared with up to 40 percent in the higher income groups, despite a similar dependency ratio.

Furthermore, the poorest households in Adele Keke worked fewer days (thereby receiving much lower food payments) than the less poor; 21 days versus 71 days, respectively. Female-headed households in Adele Keke also appear to have been disadvantaged (in terms of days worked) when compared with households headed by men. Male-headed households worked an average of 62 days a year in FFW, compared with only 17 days per year for households headed by women.

These differences may represent a selection bias toward "able-bodied members." Yet, with the exception of Dinki, dependency ratios do not differ much between the income groups. A stronger, more explicit, consideration of poverty targeting is clearly required to strengthen the food-security potential of such projects.

Food Wages

The standard food wage recommended by the Ministry of Agriculture was 3 kilogrammes of grain and 120 grammes of oil per day. This ration covered daily subsistence requirements for six people, offering some 1800 kilocalories per head (Holt 1983; Admassie and Gebre 1985; UNEPPG 1987). However, there were many deviations from the standard because of constraints on availability and difficulties in estimating numbers of potential participants (Hareide 1986; USAID 1987).

Payments did not vary greatly by gender of household head within projects. Households headed by men or women both received the average wage for their particular locality. However, wages did vary across income groups and also between projects. For example, poorest households in Dinki and

Gara Godo received payments that exceeded not only the standard payment for a day's work (an average of 3.6 kilogrammes per day at each site), but also the payments received by less poor households in the same communities (who received between 2.2 and 3 kilogrammes). By contrast, poorest households at Beke Pond and Adele Keke received average daily payments of between 0.6 and 1.3 kilogrammes, well below the recommended ration. The low grain payments in Beke Pond arose because FFW was seen by the implementing NGO as a means of initiating activities that would become self-sustaining, and because the NGO did not wish to establish a dependency relationship.

Food wages were mostly consumed by the household. In Wollo, Kohlin (1987) found that 71 percent of FFW payments were consumed. Admassie and Gebre (1985) derived an average of 85 percent from their surveys. A comparable figure of 72 percent was found in Gara Godo. The remainder was sold to purchase clothing and other durables (24 percent) or to pay taxes (4 percent). However, in Dinki and Adele Keke, sales were much lower. In Dinki, all food wages were consumed and in Adele Keke 97 percent of the wage was consumed, with 3 percent sold to purchase salt and sugar. Sales of the food wage during 1988 averaged 137 kilogrammes per household in Gara Godo despite an average price of only 0.2 birr per kilogramme received for wheat, compared with an average 1987 price of 0.5 birr per kilogramme.

Which brings us to the food wage versus cash wage debate. The evidence that participants would support a shift in labour-intensive works away from food wages to cash wages (currently much discussed in Ethiopia) is not strong. Kohlin's (1987) study found that 70 percent of respondents preferred a food wage. Admassie and Gebre (1985) found that 90 percent voted in favour of food. The present findings do not challenge earlier ones. In Gara Godo, 89 percent of respondents preferred food, and in Adele Keke the figure was 91 percent. Only in Doma was it found that 83 percent of sample respondents preferred a cash wage over food. As noted in Chapter 3, this was largely because of fears of delays in food payments resulting from the logistical difficulties of reaching isolated locations.

Given that income transfer is one of the key objectives of such projects, the more important question is: What is the real value of the alternative wage forms to the participants themselves? Holt (1983) estimates that in the early 1980s the value of a cash wage was roughly 2.5 birr, whereas the average cash value of an FFW ration at local market retail prices was at least 2.7 birr. However, few studies provide comparative data on the relative purchasing power of food versus cash wages on a time-series basis.

One exception is Erni's (1988) evaluation of a Lutheran World Federation project in Dinki. Taking the period from late 1986 to early 1988, Erni found that only during the first half of 1987 would a fixed cash wage have had a higher market value to recipients than the food wage actually received. He concludes that the food wage was more urgently needed and more popular than a cash wage would have been.

This can be confirmed for other parts of the country, but it cannot be seen as a universal principle. For example, in Arssi FFW rations would have been of greater value to recipients than cash (in terms of a maize equivalence at prevailing prices) twice during the 1980s: at the peak of the famine (1985 to late 1986) and again temporarily at the end of 1988. At other times a cash wage would have been more valuable.

However, the story is different at other sites. In Sidamo, a food wage was more valuable than a maize-equivalent cash wage from 1985 to late 1986. However, aside from a few months in 1987 the food wage remained more valuable than a cash wage up to the end of 1989. What is more, the value of food wages in Hararghe remained consistently and substantially above the value of a cash wage throughout the period from 1985 to 1989.

Although these are theoretical comparisons they reinforce the earlier conclusions that where market integration is poor, markets operate very differently from one region to the next, and food transfers via public works do have a valuable role to play in regions in which markets are the most constrained. In other words, close consideration of local conditions is essential to decisions about the most appropriate intervention at each location, and even for different phases of a crisis. A mixed food and cash wage may be preferable to single-commodity wages in order to allow greater flexibility in wage-value adjustments according to seasonal and annual fluctuations in purchasing power. The wage value itself may also be varied according to technical or social (targeting) objectives. A flexible, rather than dogmatic, approach to the design of public works is required, resting on complementary combinations of cash, food, and technical resources rather than on single resource inputs alone.

Technical and Social Issues

Food-for-work is not just about providing work; it is also about putting labour to useful effect. The long-term impact of FFW depends on the technical value of a project and on its use to local communities. Two of the factors most widely blamed for failed projects are poor technical design and implementation, and low community involvement in project selection and subsequent maintenance (Burki *et al.* 1976; Maxwell 1978; Holt 1983; Clay and Singer 1985; Bryson, Chuddy and Pines 1990).

On the technical side, design problems may result from a conflict between relief and development objectives. Most FFW activities in Ethiopia were initiated in response to crises, both food crises and those of an environmental nature (linked closely to the former). Thus, projects have often been organized in an atmosphere of urgency, with limited assessment of community needs or of the technical validity of projects. Also, many agencies have limited management capacities for rapid expansion into the field of public works.

As a result, projects list their achievements in terms of the length of roads built or how many millions of trees have been planted, but the quality and sustainability of such assets have rarely been assessed in detail. Such a system emphasizes the quantity rather than the quality of outputs. For example, the Dinki road project achieved much in a short period of time—680 kilometres of rural roads were created through 2 million workdays of labour, mostly during the first half of 1987 (Erni 1988). But it was later found that more work was required to "arrange for proper water control for less maintenance, and in some places reconstruction of basically wrong alignments" (Erni 1988).

Brown (1989a) argues that most large labour-intensive projects in Ethiopia have only basic participatory and technical planning. Vallee's (1989) evaluation of 24 projects in Hararghe concurs, noting particularly a lack of adequate planning and predictions of food payment needs. He concludes that, "without good [planning] a technical project involving food-for-work leads to total confusion."

Such assessments are often echoed by the participants. In Gara Godo, 77 percent of respondents felt that none of the assets created will be operational within 10 years. The main reasons were that bunds and terraces were often erected too hastily, often on bare hillsides, and that roads were poorly finished and unprotected against heavy rains. In Adele Keke, 30 percent of respondents claimed that most assets will be useless in 10 years. Another 35 percent qualified their response by saying that assets might survive if maintained.

But will they be maintained? It has been argued that the main problems with public works are that labour motivation, productivity, and long-term commitment all tend to be low (Maxwell 1978; Clay and Singer 1985; Drèze 1988). Part of this may be ascribed to the characteristics of the labour attracted by public works—the distressed and disadvantaged—but much may be ascribed to a lack of community participation in project selection and implementation. Detailed consultation between communities and agencies before projects start is uncommon. According to Tato (1989), "there is no involvement of the local community in the planning process." This was confirmed by respondents at the Adele Keke site, where 97 percent of households have never been asked their opinions about the project, even by on-site project managers. The record is only slightly better in Gara Godo, with 86 percent of respondents complaining of poor consultation.

The same applies to efforts spent by participants on a project and the effort put into its maintenance. If participants do not perceive personal benefits from the project, their application to the task at hand is understandably low. The negative attitude of farmers to soil bunds has been widely reported, due to immediate losses of up to 10 percent of farmland (Admassie and Gebre 1985; EEC 1989; UNSO 1991). Yet more than 70 percent of labour used in

1985 went to soil conservation (Franklin 1986; Hareide 1986). Admassie and Gebre (1985) reported negative attitudes toward water-control activities (river diversion and spring development), because continuing droughts obscure the benefits. The low survival rates of new trees is widely documented with estimates of survival 1 year after planting ranging from optimistic highs of 80–85 percent (EEC 1989; Finucane 1989), to lows of 0–15 percent (Brown 1989b; EEC 1989; Wood and Stahl 1989).

Problems in all three areas seem to stem from disagreements between participants and managers about the best species of tree to be planted where, about whom the tree (or bunds) will belong to in the long term, and whether the digging of ponds is the most appropriate project for immediate local needs. When sample respondents in Adele Keke were asked what projects were least liked, the most common responses were stone bunds, terracing, and tree planting. Only 3 percent of households believed that the assets created would benefit them personally: more than 21 percent thought that the government would own all assets and the remaining 76 percent felt that assets would belong to the community as a whole.

When asked what they would rather do instead, a majority of respondents (57 percent) desired to consolidate and extend roads that they had already created. In Gara Godo, on the other hand, pond excavation was the least popular activity, and many people would rather have built schools and clinics. Almost 41 percent of respondents believed that the assets generated would belong to the community. The remainder felt that they would belong to the government. As a result more than 56 percent of respondents were sure that the assets would never benefit households that had not participated in the project; the food wage was widely perceived to be the only direct benefit of FFW.

Improved consultation between communities and project organizers should clearly be a high priority for future projects (Brown 1989a; Diriba 1991). Without a consensus on the validity of project objectives, and a clear understanding of asset ownership rules, few households will be willing to participate wholeheartedly in, and subsequently maintain, food-for-work projects (EEC 1989; WFP 1989).

On the other hand, there is little doubt that FFW was largely an effective means of transferring income to vulnerable households in times of need, as well as the vehicle for some valuable development activities. The best resource conservation works were those that required the least maintenance— vegetated bunds and terraces on common-access (but often enclosed) hillsides used purely for tree planting. Furthermore, the multiplier effects of road building in certain regions were marked. For example, 3 years after the road reached Dinki, two water mills had been established (with parts brought in by four-wheel-drive trucks), new fruit plantations were planted and the traditional cotton spinning and weaving industry had been revived, all because of much-improved access to highland markets. Many lessons have been learned with FFW in Ethiopia, from which modified projects may benefit immensely in the future.

Cash-for-Food

The other type of public works in Ethiopia was the cash-for-food (CFF) programme. This innovative scheme assumed that in certain parts of the country it was a lack of purchasing power, rather than a lack of food, that was causing most hardship. It was therefore believed that recipients of cash in such "pockets" of famine could tap into regional markets where food was still available at reasonable prices, thereby stimulating flows of food to distressed areas. Furthermore, it was anticipated that this would reduce intervention costs (especially compared with FFW), reduce delays in wage deliveries, and prevent migration toward food distribution points, all while supporting development projects at a community level (Kumar 1985; D'Souza 1988).

Initially a flat rate of 35 birr (US$17) was paid to all households regardless of size, in addition to a lump sum of 30 birr per household designed for the purchase of seeds (UNICEF 1988). Later a sliding scale was introduced, with sums ranging from 35 to 65 birr distributed monthly according to household size. In return, recipients undertook village-based development projects. By 1988, over US$5 million had been disbursed to 120 000 households at 19 project sites in Shoa, Arssi, and Gamo Gofa (UNEPPG 1989e).

The sites surveyed were at Korodegaga and Doma. In both cases, irrigation projects were initiated by UNICEF and the RRC with a view to improving longer term food security for the communities involved. Korodegaga started receiving CFF payments in early 1985, and Doma's project began in January 1986. Cash distribution ended in mid-1988. No targeting was attempted; all households were included in CFF in both villages. According to respondents, there was no discrimination against poorer or female-headed households because payments were made strictly according to household size.

In both villages, payments were mostly used for the purchase of food. On average, 96 percent of the households in Korodegaga used their payments primarily to purchase food. As would be expected, the proportion was somewhat higher among lower income households (97 percent) than for those in the upper tercile (91 percent). Other uses were purchases of clothing and payment of local taxes. There was little difference in the use of payments by gender of household head.

In Doma, on the other hand, the proportion of households using payments only for food averaged 80 percent: 94 percent in the poorest households versus 73 percent among the less poor. Female-headed households (eight in the sample) spent all of their cash on food, whereas in male-headed households (at least those in the upper and middle income groups) the pattern was more diversified, with some households saving income for several months to purchase cloth and livestock.

However, households in Doma spending CFF income on nonfood items were an exception. In general, respondents felt that monthly payments were insufficient to cover subsistence requirements for an entire month. In Korodegaga, more than 90 percent of the poorest households could not survive for a whole month on the cash payment alone. Even less poor households were unable to stretch the payment from one month to the next. There was no difference according to gender of household head. As a result, most households attempted to earn income from other sources: 78 percent of male-headed households sold firewood, 3 percent found temporary labour in nearby towns, and the remainder survived by selling assets. The female-headed households all depended on sales of firewood. In Doma, on the other hand, only 60 percent of households complained that CFF payments were too small to last one month.

Why the difference? Not because they were less affected by the famine. On the contrary, Doma was severely affected by the crisis in 1985/86, and had not yet recovered in 1990. The explanation revolves more around two other factors that hindered the ability of households in Korodegaga to convert cash into food. First, although overt corruption was deemed insignificant at these projects, CFF payments in Korodegaga were often delayed or even interrupted over several months. As a result, many households were forced to borrow cash from money lenders against the value of the next payment, at interest rates averaging 100 percent.

The need to turn to this borrowing facility (only made available to CFF participants because lenders were willing to accept UNICEF's credit rating) reduced the net welfare effect of the distribution programme. The high interest rates severely reduced quantities of food that could be purchased by CFF recipients after each cash distribution. In Doma, on the other hand, payments were made without interruption and few households went into debt against the value of their next CFF payment.

The second, more important, problem was that cash recipients in Korodegaga had difficulty in fulfilling the expectation of tapping distant food markets and stimulating flows of food from surplus to deficit regions. During the famine, government policies restricting grain movement between localities were enforced at road control stations and by local militia. Consequently, although many households travelled to highland markets up to 100 kilometres away in the search for grain at reasonable prices, the return journey presented many hazards. More than 50 percent of respondents in Korodegaga had food confiscated on the way home. More than a dozen households had their grain confiscated five to ten times during 1985 and 1986. Some were arrested and charged with the illegal transport of grain.

As a result, many households decided to stay home and purchase food locally. This caused local inflation of food prices at the time of each cash distribution. More than 75 percent of households reported that prices rose by as much as 100 percent around distribution times. It is perhaps not surprising, therefore, that 80 percent of respondents would rather have received grain at their doorstep than face the problems associated with cash payments.

In Doma, on the other hand, conditions were different. Almost 80 percent of respondents were satisfied in receiving cash rather than food. Doma is more isolated than Korodegaga, lying 100 kilometres south of the nearest town where police are headquartered. Thus there were no constraints on CFF recipients travelling up to 2 days' walk to the market town of Kemba in search of cheaper food. No household had its grain confiscated on the journey home, and the impact of CFF distributions on local prices was therefore less than in Korodegaga (price rises rarely exceeded 30 percent).

ASSET TRANSFERS

The distribution of assets to distressed households has become an increasingly common crisis intervention in Ethiopia (Band Aid 1987; UNICEF 1988). As outlined by Jodha (1975), the rationale for this is that the "major source of growth to be protected through timely drought relief is the farmer's own production base reflected through his own capital assets, including livestock."

In large parts of Ethiopia, access to draft oxen is a prerequisite for success-ful farming (Gryseels et al. 1988). However, in 1983/84, 70 percent of house-holds did not have access to a pair of oxen. Farmers not owning their own oxen (usually the poorest households) traditionally engage in sharing, ex-change, or rental arrangements with wealthier oxen owners (McCann 1987a). Although the existence of such systems allows resource-poor farmers access to draft power (albeit only after the owners have finished their own fields), it places them in a position of dependence on wealthier farmers. Ox-distribution projects aim to help poorer farmers become more independent. Recipients of oxen on credit still need to work within an exchange system (traditional ploughs require a pair of oxen to pull them), but the exchange is carried out on a more reciprocal basis.

A project for distributing oxen on credit, as well as free seeds, ploughs, and animal feed was initiated in 26 villages of northern Shewa (including Dinki) in 1985. The proportion of households without any oxen at all in Dinki in 1985 was 61 percent (Hadgu 1985; Jutzi 1986). Roughly half of these (the pro-gramme was targeted at the poorest households) received an ox during 1985 and 1986 (Gryseels and Jutzi 1986). In 1985, oxen were purchased in the highlands and walked to the lowlands for distribution. This led to problems of climatic adaptability, with many recipients complaining that highland oxen were unsuited to the hotter lowland conditions and too weak to plough.

During 1986, it was therefore decided to give cash instead (300 birr) so that households could purchase their own oxen (McCann 1985). Recipients signed an agreement to repay the loans within 2 years from September 1987. By 1989, no household had repaid any of its loan, and the project decided not to pursue repayment.

The targeting of poorest households was largely successful in Dinki. A greater proportion of households (72 percent) receiving an ox, or the cash to purchase an ox, were among the poorest households in the community. On the other hand, there was also some mistargeting. First, some recipients already owned one or more animals. For example, 17 percent of less poor recipients were given a second, not a first, animal. Even among the poorest recipients, 9 percent were already in possession of an ox. However, given that all households justify categorization as "poor" at this extremely deprived location, such limited mistargeting is not unduly worrisome.

More important, 34 percent of households in the lowest tercile tried to be registered for an ox but were turned away. These were told that the quota was full (no oxen left) or that their registration had been lost, or they were promised an ox but never received it. Many of the latter were female-headed households. This oversight was rectified in 1986 through the distribution of cash to households headed by women, mainly for the purchase of milch cows (McCann 1985).

Unfortunately, the poorest households had more difficulty in retaining and capitalizing on their new assets than did the relatively wealthier households. Forty-five percent of the poorest recipients lost their ox soon after receiving it (sold or died), compared with only 17 percent of less poor households. In most cases, losses were due to feeding problems coupled with the poor condition of the animals received. More than one-third of all respondents complained that the ox was either in "poor" or "very bad" condition when they received it. A zebu ox in good condition weighs between 300 and 350 kilogrammes, but project oxen weighed on average less than 275 kilogrammes, and some weighed little over 200 kilogrammes (Gryseels and Jutzi 1986).

What is more, although most recipients were given 120 kilogrammes of hay and supplementary feed blocks, many found it very hard to find sufficient feed to maintain the original weight. Almost 20 percent of recipients sampled resorted to feeding thatch from their own roofs as well as cactus stems. Consequently, many animals died or had to be sold. Most of the households forced to sell their ox did so during the first harvest season, or less than three months after receiving them. The urgency of distress sales can be ascribed to the fact that immediate food needs were greater than the ability of households to wait three months until the next harvest.

Although the loss of recently acquired assets was significant for recipient households, some other households benefited from high animal losses from the project. This was because the project distorted traditional systems of oxen and labour exchange. Prior to 1985, almost 30 percent of sample households engaged in a system called *ye qollo*. In this system, one household loans an ox to another for a year in return for a pre-agreed amount of the ensuing harvest. The borrower then joins another household, which also has access to a single animal, in order to make a pair. The pair is then shared between the two farms on alternate days (McCann 1987b).

During the project, the influx of three dozen new oxen made it easier for households without animals to bargain down the price of a loan. According to respondents, the loan of an ox in 1984 cost an average of 320 kilogrammes of grain. During the project this fell to 270 kilogrammes in 1985 and as low as 180 kilogrammes in 1986 (Gryseels and Jutzi 1986). The price collapse was not due solely to the project because drought-related contraction of farmed areas also played a role. However, many respondents said that this disruption of the labour market caused animosity to develop between lenders and borrowers. However, feelings appear to have eased more recently. Because of animal mortality and sales, oxen rentals in 1989 had risen to 320 kilogrammes, placing further strains on the poorest households in their attempt to recover from famine.

In addition to the oxen, many households also received a package of 120 kilogrammes of seeds. Targeting of the poorest households was effective in this distribution, with 72 percent of recipients being poorest households, compared with only 22 percent being less poor. Yet there were three problems with this part of the project. First, although the rains had already started, a number of households consumed most of the seed immediately and had none left to plant.

Second, although most households sowed their seed on time (distribution took place in May/June of 1985), the 1985 rains were sporadic and poor. Thus, almost 20 percent of households lost their seed during dry spells in the middle of the rainy season. These households argued that several monthly distributions of seed would have been more advantageous, enabling them to catch the later rains.

The third problem was that the maize variety provided (the major portion of the seed package) was less successful than expected. Lack of germination and low yields were reported by 36 percent of recipients, most of whom argued that they were given a highland maize variety that was unsuited to the drought conditions of the lowlands.

TECHNOLOGY TRANSFERS

It is widely recognized that sustainable growth in agriculture, with its attendant improvements in food security, depends largely on technological change (Mellor, Delgado, and Blackie 1987). Such change can be of a longer term nature, such as the hybridization of cultivars to meet local requirements, or of a short-term nature. One such short-term intervention was the single-ox plough.

Linked to the Dinki ox–seed distribution considered above was a project component designed to have a more lasting effect on the farming system than the rebuilding of draft-animal stocks. Some farmers receiving the ox–seed package were also given a new type of plough that could be drawn by a single ox rather than a pair. Designed by the International Livestock Centre for Africa (ILCA), this innovation was thought to hold promise for the rapid regeneration of asset-depleted farm economies (Cross 1985). Tests in the highlands indicated that a well-fed single ox could cultivate up to 70 percent of the area normally ploughed by a pair, with no loss of overall yield (Gryseels *et al.* 1984). It was therefore assumed that farmers with access to only one ox would benefit from reduced dependency on traditional rental arrangements.

The results were disappointing. Most farmers received the plough in June 1985 when the ploughing season was already under way. This allowed no time for training in the new technology. As a result, farmers tried out the plough only during seeding, covering, and weeding. Few were satisfied. Even light tasks, such as soil covering and weeding, required much more time to complete than with a traditional pair (partly because many farmers worked their animals only 4 hours a day for fear of wearing them down). According to 72 percent of recipients the technology failed because local soils are too heavy and stony, and because the oxen were too weak to pull a plough alone. The remaining 28 percent argued that they might have adopted it had they not been forced to sell their oxen, or if the oxen had not died prematurely.

In the event, after a few trials, most households never used the plough again. Less poor households, most of whom owned a pair of oxen already, did not bother to use the plough (Table 6.4). In 1986, poorest farmers tried to use the plough more often than the less poor, but only for light tasks. Since 1986, only 23 percent of households have used the plough. Several commented that they had chopped up the beam and yoke for firewood.

Table 6.4. Households receiving and using the single-ox plough during 1985/86, by income group (source: IFPRI survey, 1989/90)

	Income groups (percentage of households)		
	Poorest	Middle	Less poor
Received plough	33	66	17
Use of plough in 1986 for:			
soil preparation	17	28	0
seed covering	0	6	0
third weeding	11	22	0
Used plough after 1986	6	17	0

Gryseels *et al.* (1988) found similar results when evaluating adoption rates for the new plough in villages near Debre Berhan. Although preliminary trials were encouraging, "the number of farmers using the system remained a minor fraction of the farm population." The main factors cited for nonadoption were again heavy soils, weak oxen, and excessively sloping land. It was concluded that further research into plough adaptation, coupled with a focus on improving the condition of draft animals, was needed.

PUBLIC INTERVENTION COSTS

Thus far in this chapter we have examined operational considerations relating to selected public interventions against famine. The discussion has shown that public response to crisis was characterized by the activation of a wide range of projects and programmes. However, a question yet to be raised is: What is the optimal mix of such interventions for achieving a cost-effective reduction in household vulnerability to famine? Although a precise answer lies beyond the scope of this report, the following section addresses the issue by outlining the key data that would be required for future cost estimations.

Each of the projects considered above works toward a common goal of protecting households against the undesirable consequences of famine (mortality, dislocation, destitution), be it in the short or the longer term. Of course, different interventions do not operate in a vacuum. The marginal welfare gains (or losses) associated with individual projects may depend on what other projects and policies are being implemented in the same location. For example, free food aid may compete with food-for-work, and the benefits of cash-for-work may be compromised by policies inhibiting private trade. Thus, a public strategy for famine intervention should ideally be based on a sound understanding of the trade-offs between projects and policies, and between alternative household activities. And, such an understanding must be firmly grounded in an analysis of the cost-effectiveness of each individual action taken.

However, quantification of intervention costs is not straightforward. The cost-effectiveness of a project designed to transfer calories or income depends not only on the marginal income transferred, but also on the subsidy component of the scheme (Reutlinger 1988). The marginal income conveyed can be calculated as the difference between income received by project participants and the cost to the participants (potential income foregone in other activities). And the subsidy component represents the difference between the value of assets or services generated and the cost of the scheme to the public.

Unfortunately, data required for analysing projects according to these parameters are rarely obtained in an operational setting (Alderman and Kennedy 1987; Berg 1987). According to Torry (1984), "next to nothing is known in detail about how satisfactorily various emergency interventions and adjustments actually work, judged by any standard measure of effectiveness, for any sample of famine-affected households and villages." Many evaluations do provide detailed audits of input disbursements (food, cash, assets), numbers of beneficiaries, and gross costs (NORAD 1984; RRC 1985b; Jareg 1987; UNHCR 1988). But, few generate the data required for detailed net cost calculations (Svedberg 1991).

For example, UNICEF (1988) attempted a cost comparison between cash-for-work and food-for-work. Although the prima-facie costs of cash-for-food are found to be comparatively low compared with food-for-work (Table 6.5), the cost accounts for both project types are no more than partial financial analyses, which exclude local and regional (private) marketing costs, net inflationary effects on local markets, opportunity costs of participants' time, the discounted present value of income received by target groups from the capital goods (assets) created by the projects, or even the sustainability of income derived from assets created.

A more complete assessment of the comparative costs of these two projects might therefore be hypothesized as follows:

$$\text{cost-effectiveness} = \frac{\text{incremental calories consumed by target group}}{\text{total costs}}$$

where incremental calories are the net calorie effect for participants, effect on prices, and effect on production, while total costs = project fiscal costs + private costs + national costs.

In such a calculation, the various project, private, and national costs to be taken into consideration might include the following.

- Project fiscal costs—food import and distribution costs (commodity costs, sea freight, storage, road transport); implementation costs (including consultants' reports); and administration.
- Private costs—opportunity cost of target groups' time (income foregone); local deflationary costs to food producers; and transport.
- National costs—government administrative costs (salaries, per diems); and depreciation of facilities.

None of the interventions considered for this study collected all of the data required. A broad indication of the overall costs associated with some of the projects considered in this chapter is given in Table 6.6. Unfortunately, even these data generally exclude fixed costs, foreign personnel costs, recurrent operational costs, depreciation of equipment and materials, opportunity costs of local labour, or total numbers of actual beneficiaries.

However, Table 6.6 does show that there are variations in project costs per beneficiary even if one compares only gross project costs. For food-aid distribution, the range runs from US$12 (in relation to the very accessible Debre Berhan area) to US$50 (for the very inaccessible villages around Dinki).

Table 6.5. Costs (US$1000) of selected public works interventions (sources: WFP 1986; Erni 1988; UNICEF 1988)

Type of intervention	Organization	Food cost	Capital	Inland transport, storage, and handling	External tranport, storage, and handling	Administration	Total project cost	Cost per year
Food-for-work	World Food Programme	48 541	–	16 244	11 180	145	139 056	46 352
	Government	–	16 946	–	–	19 652		
	Other	–	6174	20 174	–	–		
	Lutheran World Federation	2963	–	1967	–	1534		
	Other	–	67	65	–	–	6596	3298
Cash-for-food	UNICEF	–	445	6	–	7	458	458

Table 6.6. Overview of costs (US$1000/year; excluding overheads and professional personnel) related to selected project interventions (sources: Gryseels and Jutzi 1986; WFP 1986; Erni 1988; UNICEF 1988; Redd Barna records at Bolosso Sora 1989)

Project category/type	Duration	Donor cost		Government cost		Total participants (households)	Cost per participant (US$/ household)	Main donors	Implementing agencies
		Capital	Food	Capital	Food				
Public works: food-for-work	1986–1988	3500	2963	–		31 250	69	Canada/ ADC/ILCA/ Caritas	LWF
food-for-work	1988–1989	n/a	310	–		8800	(35)	Redd Barna	Redd Barna
food-for-work	1987–1990	27 568	48 540	62 946		1 000 000	46	WFP	Ministry of Agriculture
cash-for-food	1984–1988	22	431	n/a		2016	(39)[a]	UNICEF	UNICEF/ RRC
Food transfers: food aid	1985	6	46	–		597	12[b]	OXFAM (USA)/GTZ/ ILCA	ILCA/ OXFAM
food aid	1985	586	876	–		29 000	50	Canada/ ADC/ILCA/ Caritas	LWF
Income transfers: Ox–seed distribution	1985–1988	102	17[c]	–		597	50	OXFAM (USA)/	ILCA/ OXFAM/ AFCF/GTZ

n/a, not available.

[a] Based on UNICEF's evaluation of a single cash-for-food site for 1 year of operation (1987).

[b] Eight kilogrammes of wheat were distributed to each of 597 households (mean size 6.4) for 6 months. Wheat (mixed varieties) was purchased from Agricultural Marketing Corporation at wholesale price of 51.65 birr/100 kilogrammes in 1985/86 (US$1.00 = 2.05 birr).

[c] "Food" costs in this instance represent the estimated value of the seed package distributed to each participant household, based on 1985/86 Agricultural Marketing Corporation wholesale prices.

Where food-for-work is concerned, the costs vary from US$35 to US$69. The latter differences can be ascribed partly to scale of operation. Whereas one food-for-work intervention may depend on imported food and draw considerable external handling costs as a result (much of the cost would be shared with the host government), another smaller project may involve purchasing local foods, thereby avoiding external transport, shipping, and handling costs, but its relatively much larger inland handling costs would not be shared with a government partner because only a sole implementing agency would be involved.

These wide differences in gross project costs, coupled with the other serious data limitations outlined above, argue for more thorough, standardized (yet transparent) accounting procedures among agencies involved in famine work in order to permit improved cost calculations. This applies not only to the government and donors but also to NGOs.

Wolde-Mariam (1991) calls international NGOs "either a sort of transnational charity organization or transnational business, or a mixture of both." It is essential at a time of public sector reform and increasing government and donor reliance on such organizations ("because they know the people"), that NGOs be held to the same standards of cost-efficiency, transparency, and accountability that are demanded of any other organization that operates in the public domain (Fowler 1992; Riddell and Robinson, 1992; Brett 1993). Competition between NGOs over spheres of influence, their often poor record of project monitoring and impact analysis, and the common veil of secrecy that shrouds questions of cost-benefit serve to assist, perhaps least of all, the NGOs themselves. Much closer integration of NGO activities and monitoring into host-country regional development planning could resolve many of these problems.

Given limited public resources for famine alleviation, concern about cost-effectiveness is not merely academic. The choice of projects and policies used in building up an optimal package of public responses must be guided by considerations of (1) intervention coverage (are vulnerable target groups adequately reached, be it directly or indirectly?), (2) cost-efficiency (is increased food security being achieved at least cost (both fiscal and human)?), and (3) efficacy (are actions actually moving vulnerable households out of food insecurity, as opposed to just reaching them at a reasonable cost?).

An assessment of the comparative costs of intervention types and scales, based on better net-cost information as outlined above, remains essential to improved priority setting as part of an effective counter-famine strategy. Such an assessment should be a high priority as an early step of any national famine-preparedness strategy.

NOTES

1. There was often more than one intervention at each of the survey sites. The agencies implementing these projects comprised two multilateral organizations (UNICEF and World Food Programme), one international research institute (the International Livestock Centre for Africa [ILCA]), three NGOs (OXFAM America, CARE, Redd Barna [Norway]), and one church organization (Lutheran World Federation). At the first two sites (Doma and Korodegaga), the RRC was closely involved in the implementation of relief activities (in association with UNICEF) and was very helpful in facilitating the field surveys. The same was true for Ministry of Agriculture staff at Adele Keke.
2. For discussion on the issue see Stevens (1979), Jones (1989), Hopkins (1990), World Bank/WFP (1991), Ruttan (1993) and Deng and Minear (1993).

7   Finally Conquering Famine: the Way Forward

Africa is blighted by an anachronism. For thousands of years a universal human phenomenon, famine is now restricted to the African continent. Most other regions have found ways to deal with the problem; Africa has not. The greatest tragedy about famine today is not that it causes such devastation, but that it need not, and should not, be occurring at all.

We have seen in these chapters that most households in Ethiopia have survived probably the most traumatic 20-year period of the twentieth century—a period marked by two major famines, two revolutions, multiple droughts, and a steady erosion of food production and food security. Some households made it on their own, others survived with outside assistance. At the same time, the purveyors of assistance (national institutions, donors, and voluntary agencies) were themselves learning the hard way about how to help vulnerable households through troubled times, how to work together, and what may be done to prepare for the future.

This concluding chapter focuses on the latter. What can we learn from the failures of the past, and from the successes of other continents? What can Africa as a whole learn from Ethiopia? Although simple generalizations are to be avoided (because local specificities have been shown to be crucial to the famine story), a number of important conclusions can be drawn.

GOOD GOVERNANCE

A food-secure future without famine rests on three pillars: good governance, sound growth policies, and active preparedness.

Reference to governance does not imply a particular political dogma or procedure. Rather it means (1) an efficient use of resources that is accountable, (2) an allocation of resources in a transparently nondiscriminatory fashion in regional and ethnic terms, and (3) participatory planning and control of resources at a decentralized, local level.

Efficiency and accountability in resource use at a national level are important in their own right, but are essential for a productive relationship between the public and private sectors. Where and on what to spend scarce capital may be political decisions, but must be watched closely if accountability is real.

Accountability in resource use is essential to prevent economic mismanagement and to build trust among private sector operators in long-term public commitments and contracts. Accountability is also an essential basis for participatory interaction between government and its constituents in the development process; namely, rural smallholders.

Transparency in political decisions on where to spend resources (as well as where to raise taxes), is crucial to defusing many of the concerns that lead to civil conflict. Regional and ethnic strife often revolves around perceptions (real or otherwise) of discrimination by a central authority. Part of this problem can be addressed by improving the information base upon which economic decisions are made. It can also be helped by broad discussion of the rationale underlying long-term development objectives. Both are assisted by the existence of a free press (Drèze and Sen 1989). Perhaps a precursor to both, however, is a national commitment to decentralized and participatory decision-making.

Decentralization of decision making and resource control can improve the appropriateness of actions taken and local accountability. Improved popular participation in those decisions and in the resultant actions lies at the heart of successful development. Although this may not depend on any of the variants of democracy pursued by the industrial world, it does require a serious examination of the process of government and of the role of communities and individuals in government decision making. However, global experiences strongly suggest that a well-defined democratic process does assist economic growth and the removal of famine.

Ethiopia has experienced none of the above until very recently. Participation and decentralization were taken up as watchwords by the post-Mengistu Government. In practice this was largely restricted to a redrawing of internal administrative boundaries along ethnic lines with each new division being given a larger role in resource control than before. Commitments to transparency in how resources are allocated by region, to full accountability of resource use, to the establishment of imposable standards of efficiency, and to full press freedom were slow in making their appearance.

Nevertheless, popular mass participation in local government was encouraged along lines of experience gained in Tigray and Eritrea during the civil war. If this participation is truly "popular" nationwide, demands for better governance in all its forms will not be slow in coming.

POLICIES FOR GROWTH WITH A FOCUS ON AGRICULTURE

The second pillar involves sound economic growth policies. In the longer term it will be successful development (as opposed to relief) activities that eradicate hunger and the threat of famine. This requires not just economic growth, but growth that removes the roots of chronic food insecurity. National and regional governments, as well as donors, need to make strong commitments to allocate sufficient resources towards sustainable poverty alleviation in Africa.

With its huge dependence on agriculture, improved food security in Africa, as in Ethiopia, depends on a strategy of agricultural growth. Large investments in agricultural growth have already been made over many years in countries such as India and China, where famine was once a threat to life. These investments were focused on the promotion of technological change and commercialization in smallholder agriculture.

The adoption of improved technology is one of the keys to long-term famine prevention, both through its potential to raise food supply and through its related capacity to increase rural incomes through employment. Although opportunities for expanding land under cultivation have compensated for slow yield growth in the past, continued attempts to expand the cropping frontier would entail ever larger investments, accelerated deforestation and land degradation, and, ultimately, falling yields. Agricultural intensification is the logical alternative.

The scope for raising crop and animal yields is good in higher rainfall areas. We have seen that Ethiopian farms operate at very low levels of technology, yield, and output. Technological improvements, such as small-scale irrigation (as long as local priorities and technical feasibility are kept well in mind), improved seeds and fertilizer and mechanical threshing and milling, could all make significant contributions to raising land and labour productivity.

For example, Ethiopia's aggregate food output could be raised considerably through increased application of chemical fertilizer in favourable regions. Ethiopia uses less than 10 kilogrammes of fertilizer per hectare, compared with 60 kilogrammes in Zimbabwe and almost 200 kilogrammes in China (Russell and Dowswell 1993). Yields of maize, sorghum and rice can increase five-fold given adequate fertilizer and moisture. Fertilizer distribution is the one tangible action that is almost certain of results.

Irrigation to provide the moisture is widely considered to be an expensive option (Webb 1992). Yet, Ethiopia again mirrors the rest of Africa in that its irrigation potential is huge and almost untapped. Irrigation can increase and stabilize yields, expand area cultivated and increase incomes. In Zimbabwe, irrigated maize yields three-times as much as rainfed maize during drought years, five-times more during normal years (R. Meinzen-Dick, personal communication 1993). In many regions, small-scale, farmer-controlled irrigation in which water resources are appropriately priced deserves closer consideration.

Improved seeds can also make an important contribution. Improved hybrid maize plants in Zimbabwe have yielded 25 percent more than local species, while improved sorghum varieties in Niger yielded almost 80 percent more than local crops under drought conditions in 1984. Adding fertilizer and irrigation raised the productivity higher still.

The potential for gains, or at least stabilization, in livestock productivity should also not be overlooked. Applied disease research, support for functioning veterinary services and work on technologies for fodder improvement through hay-making and the cultivation of fodder legumes can all lead to reduced calf mortality and increased milk production in good years, and perhaps lower animal mortality in drought years, each of which are of great importance to poor households.

What is more, improved technology and practices should not be restricted to staple food production, even where improved calorie consumption is desired. Growth in the staple food sector and cash crop sectors are not mutually exclusive. Both depend on investment to the benefit of smallholders, appropriate market and price liberalization policies, infrastructural development, access to inputs and credit for the poor and improved tenure rights (for pastoralists as well as farmers). Thus, concern with national food security driven by domestic self-sufficiency should give way to the more realistic goal of improved household food security based on higher real income from multiple sources.

However, improvements in both food and cash crop productivity depend on two things. First, strong government and donor investment in agricultural programmes (including training and extensions) and in agricultural research. Such investments can have a high pay-off. A recent evaluation of 14 programmes of agricultural research in Africa since 1980 found positive rates of return ranging between 15 and 40 percent (Oehmke and Crawford 1992).

However, such investments in crop research and extension call for a reversal of declining trends in donor commitments to developing country agriculture, particularly in Africa. In the early 1990s, assistance by the United States of America to agriculture in developing countries was less than one half (in real terms) than it had been in 1988 (von Braun et al. 1993). A similar, although less dramatic, decline in assistance to agriculture was posted by the World Bank (a decline of 25 percent over the same period). Most other international and bilateral donors have shared in this negative trend. Future food supplies in Africa depend on a positive upturn.

Second, we need to understand how to deliver even available inputs to vulnerable households. The uptake of inputs will be highly regional in nature (concentrated in regions of higher potential), because the crop response will be highest there. Fertilizer, hybrid crops and livestock crossbreeds all perform differently at different altitudes and in different forms of household economy. Similarly, higher potential areas are where the private sector will choose to carry productive resources. In other words, individual inputs may succeed in raising total food supply, but demand for such inputs will be constrained in lesser potential areas.

In less-favoured environments, improvements may come relatively less from Green Revolution types of technology (higher yielding seeds, fertilizer use and water control), than through adequate investment on drought resistance in cultivar research, in improved low-input extension, and in supporting community soil- and water-management activities. The latter include management innovations such as weed control, tillage practices, mulching, and water-harvesting techniques, all adapted to local circumstance so as to avoid further degradation of fragile environments.

The latter is particularly important in the long term. Productivity increases must be sustainable. Growing food demands must be met without compromising the ability of the total stock of resources (both natural and human) to meet even larger demands in the future. Policies for sustaining the resource base should therefore focus on improving tenure rights based on local norms, control of agricultural expansion into marginal lands via the enforcement of such tenure rights, and control of resource mining, such as charcoal production.

Improved resource use, as well as agricultural growth, must be supported by the removal of constraints to production, including price constraints. We have shown that market dysfunction plays an important role in famine. Poorly developed infrastructure coupled with unfavourable trade and price policies contributed greatly to purchasing power collapse during the 1980s and prevented a balancing of supply and demand between deficit and surplus regions. Macro-economic distortions prevented materialization of growth potentials in the export sector, from which resources for investment in human and infrastructural development might have been derived.

Since droughts continue to hold sway over domestic food availability (the rainfall–cereal production link proving to be very strong in Ethiopia), and because not every region is affected by drought or famine simultaneously, inter-regional trade will be crucial to off-setting the impact of future food crises. This requires policies for market liberalization, but also substantial investment in rural infrastructure in order to allow freer markets to play their expected role in distributing inputs and receiving marketed surplus. Improved infrastructure and better connected markets are also required to raise the chances of income diversification among the rural poor. The latter remains one of the keys to survival and longer term asset building in crisis-prone environments.

ACTIVE PREPAREDNESS

The third pillar of a future without famine is active preparedness. There is little doubt that appropriate price and market liberalization policies will help Ethiopia to grow out of its poverty and vulnerability to exogenous crises in the long term. However, freer markets and a more favourable macro-economic environment may not be sufficient to address food insecurity in the short term. Regardless of how fast growth policies are implemented, food insecurity will continue to rise during the 1990s. This calls for a profound commitment to a coordinated intervention strategy aimed at protecting vulnerable households.

Such a strategy must encompass food *supply* stabilization issues and raised and stabilized *access* to food by the poor. The state has an important role to play in shaping the legal environment that protects the poor. The provision of infrastructure necessary for increased market participation at reduced transactions costs and well-targeted transfers of resources to vulnerable parts of the private sector are public sector functions. Thus, reforming the public sector in Africa in favour of mass privatization should not mean absolving the public sector of its responsibility for supporting household food security.

There are two main reasons for this. First, given the structural poverty of food-insecure households there is no simple, market-based solution waiting to be tapped. With few assets, low and variable income, limited human capital and limited access to credit or improved inputs, poor households cannot take advantage of the potential for private initiatives and generate the supply response that the planners expect to see. They cannot produce a large marketable surplus in response to high producer prices because their productivity potential is low, and any minor increment in food output is consumed at home to raise calorie consumption from abysmally low levels. What is more, growing urban demands for food cannot be met this way.

Second, the transport and marketing infrastructure required for the smooth flow of food, capital and labour around the country are still lacking. The costs of raising food production are rising almost as fast as the heralded benefits, and because most poor smallholders are net purchasers of food, they are affected by higher food prices perhaps more as consumers than as producers.

Active preparedness by the state means preparing for a crisis while simultaneously working to prevent it. Such a preparedness strategy has three principal components. First, the capability to record and diagnose distress signals and to alert appropriate institutions of the danger. Second, the pre-establishment of explicit targeting and intervention strategies to cover population groups most at risk. Third, the development of the local institutional capacity to organize an effective response to a crisis alert.

These three functions rely on clearly defined areas of central and local government responsibility, strong political and technical backing for legislation that supports action by government structures, and the appropriate financial backing for large-scale preparation and intervention.

The first function relates to early warning. Several African governments, United Nations agencies and bilateral donors invested heavily during the 1970s and 1980s in systems for the collection and collation of data that can assist in recognizing danger signals. Until the recent past, most systems focused on indicators of food production failure obtained through remote sensing (satellite imagery), and through analysis of national data on rainfall and crop estimates.

However, early warning is not just about satellite imagery and the calculation of indices of production. It is about understanding the causes of household vulnerability, determining how these differ in time and in space, identifying how they can be addressed, and laying the groundwork for local monitoring systems and the pre-siting of locally acceptable crisis interventions. Variability in household coping capacity, the essence of vulnerability, is widely believed to hold the key to the design of more effective famine early warning systems and appropriate interventions (Torry 1988).

Thus, recognizing, understanding and addressing such variability becomes crucial to cost-effective and efficient interventions. Only certain food-insecure households are forced into stressful income diversification (away from their farms) and resource mining for short-term risk reduction—actions that may carry high long-term economic and social costs. In other words, watching indicators of food supply may be important in so far as it impinges on income, employment, and prices, but for certain households it may miss the problem altogether.

This calls for a refinement of the methodology for selecting and weighting of indicators of household and regional distress, including a relative shift of emphasis that adds to the current focus on supply-side variables. As we have seen, droughts become dangerous when they accumulate in sequence and when nothing is done to respond to signals of purchasing power collapse. Thus, relatively greater attention must be paid to distribution (market) and demand-side variables (household income and purchasing power).

It also calls for more investment in the collection and analysis of information at local levels. The circulation of timely and accurate information about problems prevents reluctant governments from ignoring such problems. Analysis of food policies and their impact on the poor is a key part of such activities. Governments can be assisted in improving their capability for the collection and analysis of data to serve improved policymaking aimed at economic growth and enhanced food security. Information is often collected in the absence of a comprehensive analytical framework. Building up a government's capacity for food policy analysis, beginning with on-the-job cooperation and institution-building is a task in which IFPRI, FAO, and others are much involved.

The second and third functions of active preparedness involve improving the link between early warning and early response. Improvements in early warning have limited value if they are not unequivocally tied to timely response that is based on previously prepared technical and economic appraisals of potential interventions. The stabilization of food entitlements for poor households in crisis prone areas remains a high intervention priority. We should be planning for the next famine in Ethiopia or Somalia, while working to prevent its occurrence. Thus, key constraints to effective action identified during past crises, such as a lack of coordination among intervention agents, ineffective targeting according to need, and issues of cost efficiency, need to be confronted more openly.

Much was learned during the 1980s about the needs of famine-affected households and how to stabilize their food entitlements more quickly. The activities needed are many and varied. There is no single, public intervention that can alone eradicate famine. For example, food aid will be needed in Ethiopia for the foreseeable future to fill the widening gap between domestic food supply and demand. But the evidence considered from Ethiopia suggests that food resources alone have a limited impact, even for famine mitigation. Households may need emergency support, but they also need income and asset support if they are to stand on their own feet.

Thus, improved public crisis response, aimed at minimizing human and capital asset loss, needs to be based first on a combination of food and non-food resources that reduces income and productivity constraints while tackling the primary need of hunger alleviation. Food aid, like many resources currently used in Africa (food, technology or skills) can be interchanged for a variety of purposes. For example, food can be sold in urban markets and the cash can be used for programmes where food interventions would disturb rural markets.

Secondly, projects of all kinds need to be based on participatory planning aimed at improved targeting and communication between managers and relief beneficiaries, and more decentralized supervision of intervention activities. This applies, for example to employment programmes. In rural areas, the upgrading of infrastructure and containment of natural resource degradation are essential development tasks that public authorities can realize through the offer of private employment. Via the income transferred, such employment plays a vital role in supporting the purchasing power of the poor while generating assets. In remote parts of Ethiopia, the multiplier effects of new roads are significant. In other areas, much progress has been made in refining watershed protection and management techniques.

Where markets are severely constrained, food-for-work has an important role to play in transferring resources to the poor. Where markets operate more effectively, a mixed food and cash wage or a cash wage is the expressed preference of participants, since this offers more flexibility. In either situation, a more flexible approach to wage-setting over time could improve both targeting and efficiency.

As with other famine-relief projects, the success of such works depends on improved technical and participatory design, the complementing of nonfood resources with food, better communication with participants about recruitment and remuneration criteria (including greater flexibility in modes of payment), improved and decentralized management and supervision, and the integration of implementation with sound monitoring and evaluation.

Equally important, the experience of poor Ethiopian households shows that more than one intervention may be needed at a time. Despite their advantages, labour-intensive works programmes are not a panacea for hunger. A diversity of instruments is often required to multiply the positive short-term income transfer effects of labour-intensive programmes. Thus, employment programmes can be designed as one part of a food security package adapted to different regions that goes beyond reliance on a single intervention, for example public works in tandem with an asset distribution or a health intervention.

The latter are certainly needed. Epidemic disease spreads rapidly during famine and accounts directly for the mortality of huge numbers of malnourished and unvaccinated people. Investment in an extensive network of rural clinics could help minimize human losses during famine by containing epidemics and could improve the collation and analysis of child-monitoring data. The improvement of potable water supplies also remains a priority. All are crucial to long-term food security enhancement.

THE TWENTY-FIRST CENTURY

Adopting the three elements proposed above as part of a comprehensive anti-famine strategy would go a long way toward ending famine in the handful of African countries in which it remains a threat. This does not mean that other countries, even outside Africa, may not fall back at some stage to a level of food insecurity easily tipped into famine by an external event. Droughts will reoccur. Military conflict is a sad reality of human relations that can never be counted out. Economic mismanagement is commonplace. Any of the three could wipe out the investments of many years at a stroke, but only if governments have not prepared themselves in advance for such an eventuality.

Policymakers around the world, awakening late to the realization that *not* preparing against crises can be more expensive in the long term than responding to individual emergencies, are beginning to ask the right questions about appropriate resource allocations for famine prevention. Part of the answer will come when better information is available on the impact and costs of alternative policies and projects. Without good information few sound decisions can be taken about investment alternatives. The longer decisions are made on an uninformed *ad hoc* basis, the longer the threat of famine will persist.

The first challenge of the twenty-first century is to strengthen social security systems in tandem with growth-oriented policies aimed at generating employment for the poor. Programmes for asset and income-building and nutritional improvement are a central part of both, since they are a precondition for human welfare and human capital development.

Just as there is no universal manifestation of famine, there is no universal solution. The dangers of not addressing long-term rural food insecurity have been spelled out in the foregoing chapters, but other dangers lie ahead. Structural adjustment reforms, massive urbanization with growing urban food insecurity, the demobilization of huge armies, regional fragmentation of decision making, rapid population growth, growing numbers of displaced people, growing international debt, declining international investment in agriculture; all of these cast the famine problem in new light. There are, as a result, "new" groups of vulnerable people to be considered, new structural problems to be addressed by policy and project action, new roles for institutions to be considered.

However, public-sector administrative and financial capacities for coping with famine and for investing against it remain weak in most parts of Africa, especially at regional and district levels. The same is true of private capacities. Thus, the transition process from famine to prosperity and from centrally controlled to free market will be fragile for many years to come. The danger of a famine striking before the country has grown out of its poverty cannot be discounted.

This calls for strengthened cooperation between the public and the private sectors. It also requires that all policies and programmes should be based on a better understanding of the dynamics of the rural economy. A widely held belief in the uniformity of rural conditions prevents appropriate attention being paid to the distributional consequences of policy and project interventions. As a result the poorest of the poor (those first, and most, vulnerable to famine) are sometimes overlooked.

Yet, given peace through improved popular participation in governance, sustained poverty reduction through rural economic growth and a determined government readiness to signal and respond to crises where and when needed, a future without famine is most certainly possible.

References[1]

Abate, Alula and Teklu, Tesfaye 1979. *Land reform and peasant associations in Ethiopia: a case study of two widely different areas.* World Employment Programme Research Working Paper. Geneva: International Labour Organization.

ACC/SCN (Administrative Committee on Coordination/Subcommittee on Nutrition) 1992. *Second report on the world nutrition situation.* Vol. I. Geneva: United Nations.

Admassie, Yeraswork and Gebre, Solomon 1985. *Food-for-work in Ethiopia: a socioeconomic survey.* Research Report 24. Addis Ababa: Institute of Development Research, Addis Ababa University.

Africa Recovery 1991. Slow famine relief imperils millions. June: 3.

Alderman, Harold H. and Kennedy, Eileen T. 1987. *Comparative analyses of nutritional effectiveness of food subsidies and other food-related interventions.* An occasional report. Washington, DC: International Food Policy Research Institute.

Amin, S. 1976. *Unequal development.* New York, NY: Monthly Press Review.

Anteneh, Addis 1984. Trends in Sub-Saharan Africa's livestock industries. *ILCA Bulletin* **18**, 2–3.

Appleton, Judith 1987. *Drought relief in Ethiopia: Planning and management of feeding programmes.* London: Save the Children.

Appleton, Judith 1988. Nutritional status monitoring in Wollo, Ethiopia, 1982–84: An early warning system? Report to Save the Children Fund (UK), London.

Aredo, Dejene 1990. The evolution of rural development policies. In Pausewang, Siegfried, Cheru, Fantu, Brüne, Stefan and Chole, Eshetu (eds) *Ethiopia: rural development options,* London: Zed Books, pp. 49–57.

Aredo, Dejene 1993. The IDDIR: A study of an indigenous informal financial institution in Ethiopia. *Savings and Development* **17**(1), 77–89.

Asfaw, Gedion 1988. Disaster prevention and preparedness plan within the context of the Five-Year Plan. *Paper presented at the National Conference on a Disaster Prevention and Preparedness Strategy for Ethiopia,* 5–8 December, Addis Ababa.

Aylieff, John 1993. Statistical summary of food aid deliveries to Ethiopia 1977–1992. World Food Programme Food Aid Information Unit, Addis Ababa. Mimeo.

Baker, Jonathan 1990. The growth and functions of small urban centers in Ethiopia. In Baker, Jonathan (ed.) *Small-town Africa: studies in rural-urban interaction.* Uppsala: Scandinavian Institute of African Studies, pp. 209–227.

Band Aid 1987. *A review of Band Aid-funded agricultural rehabilitation projects in Sudan and Ethiopia: Summary report.* London: Band Aid.

---

[1] It should be noted that Ethiopian authors are listed here by their second name, following international convention, rather than by their first name which is the Ethiopian norm. This procedure is followed in order to allow for consistency between this and other international bibliographies.

The abbreviation E.C., which appears in several bibliographic entries, stands for Ethiopian Calendar. Ethiopia still applies the Gregorian rather than Julian calendar.

Barnhart, A. 1993. Four nations need urgent assistance, U.N. tells donors. *East Africa Update.* Inter Press Service, New York, NY, March 8.

Barry, R. G. 1979. Precipitation. In Chorley, R. J. (ed.) *Water, earth, and man: a synthesis of hydrology, geomorphology, and socioeconomic geography.* London: Methuen, pp. 113–129.

Bascom, Johnathan B. 1990. Border pastoralism in eastern Sudan. *Geographical Review,* **80**(4), 416–430.

Baulch, Robert 1987. Entitlements and the Wollo famine of 1982–1985. *Disasters* (**November 3**), 194–206.

Belete, Abenet, Dillon, John L. and Anderson, Frank M. 1991. Development of agriculture in Ethiopia since the 1975 land reform. *Agricultural Economics,* **6**(2), 159–175.

Belshaw, Deryke G. R. 1990. Food strategy formulation and development planning in Ethiopia. *IDS Bulletin,* **21**(3), 31–43.

Berg, Alan 1987. *Malnutrition: what can be done?* Baltimore, MD: Johns Hopkins University Press.

Bernus, Edmond 1980. Famines et secheresses chez les Touaregs Saheliens. *Africa,* **50**(1), 1–7.

Bhatia, B. M. 1967. *Famines in India: a study in some aspects of the economic history of India (1860–1965).* Bombay: Asia Publishing House.

Bille, Jean-Claude and Eshete, Assefa 1983. *Rangeland management and range conditions: a study in the Medecho and Did Hara areas of the effects of rangeland utilization.* Joint Ethiopian Pastoral Systems Study No. 7. Addis Ababa: International Livestock Center for Africa.

Blackhurst, Hector 1980. Ethnicity in southern Ethiopia: The general and the particular. *Africa,* **50**(1), 55–66.

Bocresion, S. 1992. The Borana of southern Ethiopia: A survey of pastoral society under stress. Report to the United Nations Emergency Prevention and Preparedness Group. Addis Ababa. Mimeo.

Bondestam, Lars 1974. People and capitalism in the northeastern lowlands of Ethiopia. *Journal of Modern African Studies,* **12**(3), 423–439.

Bondestam, Lars, Cliffe, Lionel and White, Philip 1988. *Eritrea: food and agricultural production assessment study.* Final report. Leeds: University of Leeds.

Bonfiglioli, Angelo M. 1988. Management of pastoral production in the Sahel constraints and options. In Falloux, F. and Aleki, M. (eds) *Desertification control and renewable resource management in the Sahelian and Sudanian zones of West Africa.* World Bank Paper 70. Washington, DC: World Bank, pp. 42–57.

Braun, J. von, Bouis, H., Kumar, S. and Pandya-Lorch, R. 1992. *Improving food security of the poor: concept, policy, and programs.* Washington, DC: International Food Policy Research Institute.

Braun, J. von, de Haen, H. and Blanken, J. 1991. *Commercialization of agriculture under population pressure: effects on production, consumption, and nutrition in Rwanda.* Research Report 85. Washington, DC: International Food Policy Research Institute.

Braun, J. von, Hopkins, R. F., Puetz, D. and Pandya-Lorch, R. 1993. *Aid to agriculture: reversing the decline.* Food Policy Report. Washington, DC: International Food Policy Research Institute.

Braun, J. von and Pandya-Lorch, R. (eds) 1991. *Income sources of malnourished people in rural areas: microlevel information and policy implications.* Working Papers on Commercialization of Agriculture and Nutrition 5. Washington, DC: International Food Policy Research Institute.

Braun, J. von, Puetz, D. and Webb, P. 1989. *Irrigation technology and commercialization in The Gambia: effects on income and nutrition.* Research Report 75. Washington, DC: International Food Policy Research Institute.

Braun, J. von, Teklu, Tesfaye and Webb, P. 1991. *Labor-intensive public works for food security: Experience in Africa.* Working Paper on Food Subsidies No. 6. Washington, DC: International Food Policy Research Institute.

Brett, E. A. 1993. Voluntary agencies as development organizations: Theorizing the problem of efficiency and accountability. *Development and Change*, **24**(2), 269–303.

Brokken, Ray F. and Williams, Timothy O. 1990. *Economic considerations for smallholder cattle, milk, and meat production and marketing: supporting institutions, marketing, and demand.* ALPAN Network Paper 26. Addis Ababa: International Livestock Center for Africa.

Brown, Jane B. 1989a. Soil conservation, forestry, and food aid in Ethiopia: some experiences and some current problems. *Paper presented at the 6th International Soil Conservation Conference*, Nairobi, 6–18 November.

Brown, Jane B. 1989b. Impact of sustainability study of WFP-assisted project Ethiopia 2488 (Exp. II). In *Proceedings of the Food-For-Work Workshop, May 25, 1989.* Addis Ababa: Christian Relief and Development Association, pp. 17–19.

Brüne, Stefan 1990. The agricultural sector: structure, performance, and issues (1974–1988). In Pausewang, Siegfried, Cheru, Fantu, Brüne, Stefan, and Chole, Eshetu (eds) *Ethiopia: rural development options.* London: Zed Books, pp. 15–29.

Bryson, J. C., Chuddy, J. P. and Pines, J. M. 1990. *Food-For-Work: a review of the 1980s with recommendations for the 1990s.* Draft report for USAID. Cambridge, MA: WPI Inc.

Bryson, R. A. and Murray, T. J. 1977. *Climates of hunger.* Madison, WI: University of Wisconsin Press.

Buchanan, Robert L. 1990. Long-term development aid to Ethiopia: OXFAM America's experience. Statement before the Joint Economic Committee of the U.S. Congress, February 27.

Burki, S. J., Davies, D. G., Hook, R. H. and Thomas, J. W. 1976. *Public works programs in developing countries: a comparative analysis.* Staff Working Paper 224. Washington, DC: World Bank.

Bush, R. 1985. Explaining Africa's famine. *Social Studies Review*, **2**, 2–8.

Caldwell, Richard M. 1992. *Ethiopia: a country profile for famine mitigation planning and implementation.* Famine Mitigation Country Profile. Washington, DC: Agency for International Development, Bureau for Food and Humanitarian Assistance, Office of U.S. Foreign Disaster Assistance.

Campbell, David J. 1990. Strategies for coping with severe food deficits: a review of the literature. *Food and Foodways*, **4**(2), 143–162.

CDS (Centre for Development Studies) 1992. *Eritrea 1991: a needs assessment study.* Leeds: University of Leeds.

Cekan, Jindra 1990. Traditional coping strategies during the process of famine in Sub-Saharan Africa. Master's thesis, Fletcher School of Law and Diplomacy, Tufts School of Nutrition, Medford, MA.

Chambers, R. 1989. Vulnerability, coping, and policy. *IDS Bulletin*, **20**(2), 651–681.

Chole, Eshetu 1989. The impact of war on the Ethiopian economy. *Paper presented at the Fourth Annual Conference on the Horn of Africa.* City College of the City University of New York, 26–28 May, 1989.

Chole, Eshetu 1990. Agriculture and surplus extraction. In Pausewang, Siegfried, Cheru, Fantu, Brüne, Stefan, and Chole, Eshetu (eds) *Ethiopia: rural development options.* London: Zed Books, pp. 88–99.

Christensen, G. 1991. *Towards food security in the Horn of Africa: the private sector in domestic food markets.* Working Paper No. 4. Food Studies Group, International Development Centre. Oxford: Oxford University Press.

Churchill, W. S. 1956. *The birth of Britain.* London: Dodd, Mead and Company.

Clark, Lance 1986. *Early warning case study: the 1984–85 influx of Tigrayans into eastern Sudan.* Working Paper No. 2. Washington, DC: Refugee Policy Group.

Clark University 1987. The Africa drought and famine, 1981–1986: Chronologies of Ethiopia, Sudan, Mozambique, Mali, Kenya, and Botswana. Report to the United States Agency for International Development. Clark University, Worcester, MA. Mimeo.

Clausewitz, C. von. 1832 (1968). *On war.* London: Penguin Classics.

Clay, E. 1991. Famine, food insecurity, poverty, and public action. *Development Policy Review*, **9**, 307–312.

Clay, E. J. and Singer, H. W. 1985. *Food aid and development: issues and evidence.* Occasional Paper No. 3. Rome: World Food Programme.

Clay, J. and Holcomb, B. K. 1985. *Politics and the Ethiopian famine 1984–1985.* Cambridge, MA: Cultural Survival.

Cliffe, L. 1989. The impact of war and the response to it in different agrarian systems in Eritrea. *Development and Change*, **20**(3), 373–400.

Cliffe, L. and Davidson, B. (eds) 1988. *The long struggle of Eritrea for independence and construction of peace.* Trenton, NJ: Red Sea Press.

Cohen, John M., Goldsmith, Arthur A. and Mellor, John W. 1976a. Rural development issues following Ethiopian land reform. *Africa Today* **23**, 7–28.

Cohen, John M. 1976b. *Revolution and land reform in Ethiopia: peasant associations, local government, and rural development.* Rural Development Occasional Paper No. 6. Ithaca, NY: Cornell University Press.

Cohen, John M. and Isaksson, Nils-Ivar 1987a. *Villagization in the Arssi region of Ethiopia.* SIDA consultancy report. Rural Development Studies No. 19. Uppsala, Sweden: Swedish University of Agricultural Sciences.

Cohen, John, M. and Isaksson, Nils-Ivar 1987b. Villagization in Ethiopia's Arssi region. *Journal of Modern African Studies*, **25**(3), 435–465.

Cohen, John M. and Weintraub, Don 1975. *Land and peasants in imperial Ethiopia: the social background to a revolution.* Assen: Van Gorcum & Co.

Collinson, M. 1987. Eastern and Southern Africa. In Mellor, J. W., Delgado, C. L. and Blackie, M. J. (eds) *Accelerating food production in sub-Saharan Africa.* Baltimore, MD: Johns Hopkins University Press.

Constable, M. 1984. *The degradation of resources and an evaluation of actions to combat it.* Ethiopian highlands reclamation study. Working Paper 19. Addis Ababa: Ethiopian Ministry of Agriculture/Food and Agriculture Organization of the United Nations.

Constable, M. 1985. *Ethiopian highlands reclamation study: summary.* Addis Ababa: Ethiopian Ministry of Agriculture/Food and Agriculture Organization of the United Nations.

Coppock, D. Layne 1991. Hay making by pastoral women for improved calf management in Ethiopia: Labor requirements, opportunity costs, and feasibility of intervention. *Journal for Farming Systems Research-Extension*, **2**(3), 51–68.

Coppock, D. Layne 1993. The Borana Plateau of southern Ethiopia: synthesis of pastoral research, development, and change 1980–91. Systems Study No. 5. Addis Ababa: International Livestock Centre for Africa.

Coppock, D. Layne, and Reed, Jess D. 1992. Cultivated and native browse legumes as calf supplements in Ethiopia. *Journal of Range Management*, **45**(3), 231–238.

Corbett, Jane E. M. 1988. Famine and household coping strategies. *World Development*, **16**(9), 1099–1112.

Cossins, N. J. and Upton, M. 1987. The Borana pastoral system of Southern Ethiopia. *Agricultural Systems*, **25**(3), 199–218.

CRDA (Christian Relief and Development Association) 1991. *CRDA NEWS: Horn of Africa Bulletin*, **3**(6), 6.

Cross, Michael 1985. Waiting and hoping for the big rains. *New Scientist*, **February 14**.

Currey, B. 1992. Is famine a discrete event? *Disasters*, **16**(2), 138–144.

Curtis, Donald, Hubbard, Michael and Shepherd, Andrew (eds) 1988. *Preventing famine: policies and prospects for Africa*. London: Routledge.

Cutler, Peter 1984. Famine forecasting: Prices and peasant behavior in northern Ethiopia. *Disasters*, **8**(1), 48–56.

Cutler, Peter 1985. *The use of economic and social information in famine prediction and response*. London: London School of Hygiene and Tropical Medicine.

Cutler, Peter and Stephenson, R. 1984. *The state of food emergency preparedness in Ethiopia*. London: International Disaster Institute.

Dahl, Gudrun and Hjort, Anders 1979. *Pastoral change and the role of drought*. SAREC Report No. R2. Stockholm, Sweden: Swedish Agency for Research Cooperation with Developing Countries.

Davidson, B. 1993. The land of lost horizons. *Manchester Guardian Weekly*, **July 11**, 25.

Davies, Susanna 1993. Are coping strategies a cop-out? *IDS Bulletin*, **24**(4), 60–72.

Degefu, Workneh 1987. Some aspects of meteorological drought in Ethiopia. In Glantz, M. (ed.) *Drought and hunger in Africa: denying famine a future*. Cambridge: Cambridge University Press, pp. 23–26.

Degefu, Workneh 1988. Climate-related hazards: its monitoring and mitigation. *Paper presented at the National Conference on a Disaster Prevention and Preparedness Strategy for Ethiopia*, 5–8 December, Addis Ababa.

Dejene, Alemeneh 1990. *Environment, famine, and politics in Ethiopia: a view from the village*. Boulder, CO: Lynne Rienner.

Demissie, Mesfin 1986. Drought and famine. *World Health*, **August/September**: 24–26.

Deng, F. and Minear, L. 1993. *The challenge of famine relief*. Washington, DC: Brookings Institution.

Desta, Engdawork 1990. The makeup and breakup of agricultural producer cooperatives in Ethiopia. International Food Policy Research Institute, Washington, DC. Mimeo.

Devereux, Stephen 1988. Entitlements, availability, and famine: a revisionist view of Wollo, 1972–74. *Food Policy*, **13**(3), 270–282.

Devres, Inc. 1986. Evaluation of the African Emergency Food Assistance Program 1984–85: Mali. Washington, DC: Bureau for Food for Peace and Voluntary Assistance, United States Agency for International Development.

Diakosavvas, Dimitris 1989. *Government expenditure on agriculture and agricultural performance in developing countries*. Discussion Paper No. 3. Bradford: University of Bradford.

Diriba, Getachew 1991. Implementing 'food-for-development': a baseline study of household food security. Report prepared for the World Food Programme, Addis Ababa. Mimeo.

Dirks, R. 1980. Social response during severe food shortages and famine. *Current Anthropology*, **21**(1), 21–44.

Donaldson, Timothy J. 1986. Pastoralism and drought: a case study of the Borana of southern Ethiopia. MPhil thesis, Faculty of Agriculture and Food, University of Reading, Reading.

Downing, Thomas E. 1990. *Assessing socioeconomic vulnerability to famine: frameworks, concepts, and applications*. Working Paper No. 2. Arlington, VA: Famine Early Warning System Project, Tulane University/Pragma Corporation.

Drèze, J. 1988. *Famine prevention in India*. Development Economics Paper 3. London: London School of Economics.

Drèze, J. and Sen, A. 1989. *Hunger and public action*. Oxford: Clarendon Press.

Drèze, J. and Sen, A. (eds) 1990. *The political economy of hunger*. vol. 2, *Famine and prevention*. Oxford: Clarendon Press.

D'Souza, F. 1988. Famine: Social security and an analysis of vulnerability. In Harrison, G.A. (ed.) *Famine.* Oxford: Oxford University Press, pp. 1–56.

EEC (European Economic Community) 1989. *Evaluation of food-for-work programs in Eritrea and Tigray.* Report to EEC by Environmental Resources Limited, Addis Ababa.

Eliassen, I. and Eriksson, E. 1991. Startvansker for folkestyre. *Stavanger Altenblad,* **December 2,** 1.

Elizabeth, K. 1988. *From disaster relief to development: the experience of the Ethiopian Red Cross.* Geneva: Institut Henry Dunant.

Ellis, J. E. and Swift, D. M. 1988. Stability of African pastoral ecosystems: Alternate paradigms and implications for development. *Journal of Range Management,* **41**(6), 450–59.

Erni, T. 1988. Report on N. Shewa relief and soil and water conservation project in Shewa, Ethiopia, June 1985–June 1988. Lutheran World Federation, Addis Ababa. Mimeo.

Erni, T. 1989. Preliminary final report on N. Shewa relief and soil and water conservation project in Tegulet and Bulga Awraja, Shewa, Ethiopia. Lutheran World Federation, Addis Ababa. Mimeo.

Ethiopia-CSA (Central Statistics Authority) 1987a. *Agricultural sample survey 1986/87 (1979 E.C.): results of area and production by sector.* Addis Ababa.

Ethiopia-CSA (Central Statistics Authority) 1987b. *Report on the current crops, weather, and food situation.* Food Information System Project. Addis Ababa.

Ethiopia-CSA (Central Statistics Authority) 1987c. *Time series data on area, production, and yield of major crops, 1979/80–1985/86 (1979 E.C.).* Addis Ababa.

Ethiopia-CSA (Central Statistics Authority) 1988a. *Crop production forecast for Ethiopia in 1987/88 (1980 E.C.): results on area and production of main crops (private peasant holdings, main season).* Addis Ababa.

Ethiopia-CSA (Central Statistics Authority) 1988b. Report on 1987/88 crops, weather, and food situation. Food Information Systems Project, Addis Ababa. Mimeo.

Ethiopia-CSA (Central Statistics Authority) 1989a. *Agricultural sample survey 1986/87: report on livestock and poultry.* Addis Ababa.

Ethiopia-CSA (Central Statistics Authority) 1989b. *Agricultural sample survey 1987/88 (1980 E.C.): results on area, production, and yield of major crops by sector and season.* Addis Ababa.

Ethiopia-ENI (Ethiopian Nutrition Institute) 1974. *Guraghe Survey preliminary report.* Addis Ababa.

Ethiopia-MOA (Ministry of Agriculture). 1984. *General agricultural survey: preliminary report 1983/84,* vol. 1. Addis Ababa.

Ethiopia-MOA/FAO (Ministry of Agriculture)/(Food and Agriculture Organization of the United Nations) 1984. *Ethiopian Highlands reclamation study,* two volumes. Addis Ababa.

Ethiopia, Ministry of Planning and Economic Development 1992. Study on social dimensions of adjustment in Ethiopia. Addis Ababa. Mimeo.

Ethiopia, National Bank 1988. *Quarterly Bulletin,* **1987/88,** No. 4. Addis Ababa.

Ethiopia, National Bank 1989. *Quarterly Bulletin,* **1988/89,** No. 4. Addis Ababa.

Ethiopia-ONCCP (Office of the National Committee for Central Planning) 1986. Workshop on food-for-work in Ethiopia. *Proceedings of the workshop on food-for-work in Ethiopia,* 25–26 July, Addis Ababa.

Ethiopia-ONCCP 1989. *A national disaster prevention and preparedness strategy for Ethiopia.* Addis Ababa.

Ethiopia-OPHCC (Office of the Population and Housing Census Commission) 1991. *Population and housing census 1984: analytical report at national level.* Addis Ababa:

Office of the Population and Housing Census Commission, Central Statistics Authority.

Ethiopia-TG (Transitional Government) 1992. Social dimensions of adjustment. Study Group Report, Ministry of Planning and Economic Development, Addis Ababa. Mimeo.

FAO (Food and Agriculture Organization of the United Nations) 1992. *Food outlook.* February. Rome.

FAO (Food and Agricultural Organization of the United Nations) Investment Centre 1990. Evaluation de la situation nationale de securite alimentaire et problemes a resoudre. Project GCPS/NER/031/NOR. Rome: FAO. Mimeo.

FAO (Food and Agricultural Organization of the United Nations) 1992. *Sustainable developments in famine-prone areas: approaches and issues.* Staff Working Paper 9. Technical Advisory Division. Rome: International Fund for Agricultural Development.

FEWS (Famine Early Warning System Project) 1991. How close to famine? *FEWS Bulletin,* **5**, 2.

FEWS (Famine Early Warning System Project) 1992. *Harvest assessment.* Report prepared for the US Agency for International Development. Washington, DC: FEWS Project, Tulane/Pragma Group.

FEWS (Famine Early Warning System Project) 1993a. *FEWS Bulletin* **2**. Report prepared for the US Agency for International Development. Washington, DC: FEWS Project, Tulane/Pragma Group.

FEWS (Famine Early Warning System Project) 1993b. *Vulnerability assessment* **June**. Report to the United States Agency for International Development. Washington, DC: Tulane University/Pragma Corporation.

Field, J. O. (ed.) 1993. *The challenge of famine: recent experience, lessons learned.* West Hartford, CT: Kumarian Press.

Finucane, John 1989. CONCERN and its experience with a food-for-work project. In *Proceedings of the Food-For-Work Workshop,* May 25. Addis Ababa: Christian Relief and Development Association, pp. 9–14.

Foster, P. 1992. *The world food problem: tackling the causes of undernutrition in the Third World.* Boulder, CO: Lynne Rienner Publishers.

Fowler, A. 1992. Distant obligations: Speculations on NGO funding and the global market. *Review of African Political Economy,* **55**, 9–29.

Franke, R. W. and Chasin, B. H. 1980. *Seeds of famine: ecological destruction and the development dilemma in the West African Sahel.* Montclair, NJ: Allanheld and Osmun.

Frankenberger, T. R. 1991. Indicators and data collection methods for assessing household food security. Office of Arid Land Studies, University of Arizona, Tucson, AR, Mimeo.

Franklin, Tom 1986. NGO lessons learned: Preliminary findings. Confidential report of the United Nations Office for Emergency Operation in Ethiopia, June 13. Mimeo.

Franzel, S., Colburn, F. and Degu, G. 1989. Grain marketing rgulations: impact on peasant production in Ethiopia. *Food Policy,* **14**(4), 347–358.

Franzel, S. and van Houten, H. (eds) 1992. *Research with farmers: lessons for Ethiopia.* Wallingford: CAB International for the Institute of Agricultural Research, Ethiopia.

Fraser, C. 1988. *Lifelines for Africa still in peril and distress.* London: Hutchinson Education.

Gamaledinn, Maknun 1987. State policy and famine in the Awash Valley of Ethiopia: the lessons for conservation. In Anderson, D. and Grove, R. (eds) *Conservation in Africa: people, policies, and practice.* Cambridge: Cambridge University Press, pp. 327–344.

Gebre-Mariam, Ayele and Fida, Gossaye 1982. *Organization and management of ponds in Southern Rangelands Development Project.* JEPSS Report. Addis Ababa: International Livestock Centre for Africa/Rangelands Development Programme.

Gebre-Medhin, Mehari and Vahlquist, B. 1977. Famine in Ethiopia—the period 1973–75. *Nutrition Reviews*, **35**(8), 194–202.

Gedion, Asfaw 1988. Disaster prevention and preparedness plan. *In the National Conference on a Disaster Prevention and Preparedness Strategy for Ethiopia*, 5–8 December, Addis Ababa.

Ghose, Ajit K. 1985. Transforming feudal agriculture: agrarian change in Ethiopia since 1974. *Journal of Development Studies*, **22**(1), 127–149.

Gibbon, E. 1987 (reprinted 1985). *The decline and fall of the Roman empire.* London: Penguin Classics.

Gill, P. 1986. *A year in the death of African politics: bureaucracy and the famine.* London: Paladin Grafton Books.

Girgre, Aklu 1991. Agricultural policy reform in Ethiopia, 1974 to 1989. International Food Policy Research Institute, Washington, DC. Mimeo.

Gizaw, Berhane 1988. Drought and famine in Ethiopia. In Zein, Zein Ahmed and Kloos, Helmut (eds) *The ecology of health and disease in Ethiopia.* Addis Ababa: Ministry of Health.

Glantz, M. (ed.) 1977. *Desertification: environmental degradation in and around arid lands.* Boulder, CO: Westview Press.

Göricke, Fred V. 1979. *Social and political factors influencing the application of land reform measures in Ethiopia.* Research Centre for International Agrarian Development. Saarbrücken, Germany: Verlag Brietenbach.

Göricke, Fred V. 1989. *Agrarian reform and institution building in Ethiopia.* Diskussionsschriften No. 16. Heidelberg: Lehrstuhl für Internationale Entwicklungs und Agrarpolitik.

Goyder, H. and Goyder, C. 1988. Case studies of famine: Ethiopia. In Curtis, D., Hubbard, M. and Shepherd, A. (eds) *Preventing famine: policies and prospects for Africa.* New York: Routledge, pp. 73–110.

Grandin, B. E. 1987. Pastoral culture and range management: recent lessons from Maasailand. *ILCA Bulletin*, **28**, 7–13.

Gryseels, Guido and Anderson, Frank 1983. *Research on farm and livestock productivity in the Ethiopian highlands: initial results 1977–1980.* ILCA Research Report No. 4. Addis Ababa: International Livestock Centre for Africa.

Gryseels, Guido, Anderson, Frank, Assamenew, Getachew, Misgina, Abebe, Astatke, Abiye and Wolde-Mariam, Woldeab 1984. The use of single oxen for crop cultivation in Ethiopia. *ILCA Bulletin*, **18**, 5–7.

Gryseels, Guido, Anderson, Frank, Assamenew, Getachew, Misgina, Abebe, Astatke, Abiye and Wolde-Mariam, Woldeab 1988. *Role of livestock on mixed smallholder farms in the Ethiopian highlands: a case study from the Baso and Worena Woredda near Debre Berhan.* Addis Ababa: International Livestock Centre for Africa.

Gryseels, Guido and Jutzi, Samuel 1986. *Regenerating farming systems after drought: ILCA's ox/seed project, 1985 results.* Addis Ababa: International Livestock Centre for Africa.

Gutu, Samia Zekaria, Lambert, Rachel and Maxwell, Simon 1990. *Cereal, pulse, and oilseed balance sheet analysis for Ethiopia 1979–1989.* Brighton: Institute of Development Studies.

Habte-Wold, D. and Maxwell, S. 1992. Vulnerability profiles and risk mapping in Ethiopia. Draft report to the Food and Nutrition Unit, Ministry of Planning and Economic Development. Addis Ababa. Mimeo.

Hadgu, Kassaye 1985. Report on the livestock condition: A sample survey of Zoma and Dinki peasant association of Ankober Woreda, Tegulet, and Bulga Awraja Shoa Administrative Region. Addis Ababa University. Mimeo.

de Haen, H. 1993. Livestock and the environment: interactions and policy implications. *Quarterly Journal of International Agriculture*, **32**(1), 122–130.

Harbeson, J. W. 1990. Statement prepared for the Joint Economic Committee Hearing on 'Sustainable Agricultural Development in Ethiopia', United States Congress, 27 February, Washington, DC.

Hareide, Dag 1986. Food-for-work in Ethiopia. *Paper presented at the workshop on food-for-work in Ethiopia*, 25–26 July, Addis Ababa.

Harrison, G. A. (ed.) 1988. *Famine*. Biosocial Society Series No. 1. Oxford: Oxford Science Publications.

Hay, R. W. 1986. Food aid and relief-development strategies. *Disasters*, **10**(4), 273–287.

Hendrie, B. 1989. Cross-border relief operations in Eritrea and Tigray. *Disasters*, **13**, 351–357.

Henricksen, Barry, L. and Durkin, J. W. 1985. Moisture availability, cropping period, and the prospects for early warning of famine in Ethiopia. *International Livestock Centre for Africa Bulletin*, **21**.

Henze, Paul 1984. Arming the Horn 1960–1980: military expenditures, arms imports, and military aid in Ethiopia, Kenya, Somalia, and Sudan. In Rubenson, S. (ed.) *Proceedings of the Seventh International Conference of Ethiopian Studies*. Addis Ababa: Institute of Development Studies, pp. 177–183.

Hindle, Robert 1988. The international community and disaster preparedness in Ethiopia. *Paper presented at the National Conference on a Disaster Prevention and Preparedness Strategy for Ethiopia*, 5–8 December, Addis Ababa.

Hoben, Allan 1973. *Land tenure among the Amhara of Ethiopia: the dynamics of cognatic descent*. Chicago, IL: University of Chicago Press.

Hogg, Richard 1985. *Restocking pastoralists in Kenya: a strategy for relief and rehabilitation*. Pastoral Development Network Paper 19c. London: Overseas Development Institute.

Hogg, Richard 1992. Should pastoralism continue as a way of life? *Disasters*, **16**(2), 31–37.

Holcomb, B. K. and S. Ibassa 1990. *The invention of Ethiopia: the making of a dependent colonial state in northeast Africa*. Trenton, NJ: Red Sea Press.

Holden, Sarah J., Coppock, D. Layne and Assefa, Mulugeta 1992. Pastoral dairy marketing and household wealth interactions and their implications for calves and humans in Ethiopia. *Human Ecology*, **19**(1), 35–59.

Holmberg, Johan 1977. *Grain marketing and land reform in Ethiopia: an analysis of the marketing and pricing of food grains in 1976 after the land reform*. Research Report 41. Uppsala: Scandinavian Institute of African Studies.

Holt, Julius F. J. 1983. Ethiopia: Food for work or food for relief. *Food Policy*, **8**(3), 187–201.

Holt, J., Seaman, J. and Rivers, J. 1975. The Ethiopian famine of 1973–4 in Harerge Province. *Proceedings of Nutrition Society*, **34**, 115A–116A.

Hopkins, Raymond F. 1990. Increasing food aid: prospects for the 1990s. *Food Policy*, **15**(4), 319–327.

Horn of Africa Report 1990. World Bank support continues in Ethiopia. *Report*, **1**(2), 3–4.

Huffman, Sandra L. 1990. *Maternal malnutrition and breast-feeding: is there really a choice for policymakers?* Washington, DC: Center to Prevent Childhood Malnutrition.

Huffnagel, H. P. 1961. *Agriculture in Ethiopia*. Rome: Food and Agriculture Organization of the United Nations.

Hurni, Hans 1988. Ecological issues in the creation of famine in Ethiopia. *Paper presented at the National Conference on a Disaster Prevention and Preparedness Strategy for Ethiopia*, 5–8 December, Addis Ababa.

Hutchinson, R. A. (ed.) 1991. *Fighting for survival: insecurity, people and the environment in the Horn of Africa.* Based on a study by B. C. Spooner and N. Walsh. Gland, Switzerland: International Union for the Conservation of Nature.

IDI (International Disaster Institute) 1983. Drought and famine relief in Ethiopia. *Disasters*, **7**(3), 164–168.

Idusogie, E. O. 1987. A review of food and nutrition situation and the relation of agricultural strategies to food and nutrition improvement policy and programs in Ethiopia. Food and Agricultural Organization of the United Nations, Nairobi. Mimeo.

IFAD (International Fund for Agricultural Development) 1989. *Special programming mission to Ethiopia: main report.* Rome.

IGADD (Intergovernmental Authority on Drought and Development) 1990. Food security strategy study. Report to IGADD by the Institute of Development Studies, Development Administration Group and Food Studies Group, Djibouti. Mimeo.

IHD (Institute on Hunger and Development) 1992. *Hunger 1993: uprooted people.* Third Annual Report on the State of World Hunger. Washington, DC: Institute on Hunger and Development.

IHD (Institute on Hunger and Development) 1993. *Hunger 1994.* Fourth Annual Report on the State of World Hunger. Washington, DC: Institute on Hunger and Development.

ILCA (International Livestock Centre for Africa) 1991. *A handbook of African livestock statistics.* Working Document No. 15. Addis Ababa.

Iliffe, John 1987. *The African poor: a history.* Cambridge: Cambridge University Press.

Interaction 1993. The relief-development continuum. Interaction, Washington, DC. Mimeo.

Intertect 1986. An assessment of 1986 food needs for the Catholic Relief Services Food Distribution Network—Ethiopia. Report on the study conducted for Catholic Relief Services. Baltimore, MD: Catholic Relief Services.

Irvine, F. R. 1952. Supplementary and emergency food plants of West Africa. *Economic Botany*, **6**(1), 23–40.

IUCN (International Union for Conservation of Nature and Natural Resources) 1989. *The IUCN Sahel studies.* Geneva: IUCN/Norad.

Jahnke, Hans E. 1982. *Livestock production systems and livestock development in tropical Africa.* Kiel: Kieler Wissenschaftsverlag Vauk.

James, John (ed.) 1989. Relief infrastructure study of Ethiopia. Draft report to UNDP/UNEPPG/WFP. Addis Ababa.

Jansson, Kurt, Harris, M. and Penrose, M. 1987. *The Ethiopian famine.* London: Zed Press.

Jareg, Pal (ed.) 1987. *Lessons learnt from relief work: Ethiopia.* Addis Ababa: Redd Barna.

Jelliffe, D. C. and Jelliffe, E. F. P. 1971. The effects of starvation on the function of the family and of society. In Blix, G., Hofvander, Y and Valquist, B. (eds) *Famine. A symposium dealing with nutrition and relief operations in time of disaster.* Swedish Nutrition Foundation, Symposium IX. Upsalla: Almquist and Wiksells.

Jodha, N. S. 1975. Famine and famine policies: Some empirical evidence. *Economic and Political Weekly*, **10**(41), 1609–1623.

Jones, Stephen 1989. *The impact of food aid on food markets in sub-Saharan Africa.* Food Studies Group Working Paper No. 1. Queen Elizabeth House. Oxford: Oxford University.

Jutzi, Samuel 1986. Post drought recovery project, Ankober Area, Ethiopia. International Livestock Centre for Africa, Addis Ababa. Mimeo.

Kaplan, R. 1988. *Surrender or starve: the wars behind the famine*. Boulder, CO: Westview Press.

Kates, R. W. 1993. Ending deaths from famine: the opportunity in Somalia. *New England Journal of Medicine*, **327**(14), 1055–1057.

Kates, R. W. and Millman, S. 1990. Toward understanding hunger. In Newman, L. (ed.) *Hunger in History*. Oxford: Blackwell.

Keller, E. J. 1988. *Revolutionary Ethiopia: From empire to people's republic*. Indianapolis, IN: Indiana University Press.

Keller, E. J. 1992. Drought, war, and the politics of famine in Ethiopia and Eritrea. *Journal of Modern African Studies*, **30**(4), 609–24.

Kelly, C. and Taylor-Powell, E. (eds) 1992. *Perceptions of famine and food insecurity in rural Niger*. USAID Working Papers No. 1. Niamey, Niger: United States Agency for International Development.

Kelly, Marion 1987. Wollo nutrition fieldwork programme. A report. Save the Children Fund (UK), Dessie. Mimeo.

Kennedy, Eileen 1992. The impact of drought on production, consumption, and nutrition in southwestern Kenya. *Disasters*, **16**(1), 9–18.

Kennedy, Eileen T. and Cogill, Bruce 1987. *Income and nutritional effects of the commercialization of agriculture in southwestern Kenya*. Research Report 63. Washington, DC: International Food Policy Research Institute.

Kidane, Asmerom 1989. Demographic consequences of the 1984–1985 Ethiopian famine. *Demography*, **26**(3), 515–522.

Kifle, Henock 1972. *Investigations on mechanized farming and its effects on peasant agriculture*. CADU Publication 74. Asella, Ethiopia: Chilalo Agricultural Development Unit.

Kloos, Helmut 1982. Development, drought, and famine in the Awash Valley of Ethiopia. *African Study Review*, **25**, 21–48.

Kloos, Helmut 1991a. Health impacts of war in Ethiopia. *Disasters*, **16**(4), 347–354.

Kloos, Helmut 1991b. Peasant immigration, development, and food production in Ethiopia. *The Geographical Journal*, **157**(3), 295–306.

Kloos, Helmut and Zein, Z. Ahmed 1988. *Health and disease in Ethiopia: a guide to the literature, 1940–1985*. Addis Ababa: Ministry of Health.

Koehn, Peter 1979. Ethiopia: Famine, food production, and changes in the legal order. *African Studies Review*, **22**(1), 51–71.

Kohlin, Gunnar 1987. Disaster prevention in Wollo: the effects of food-for-work. Report sponsored by Swedish International Development Authority, Stockholm. Mimeo.

Kumar, B. G. 1985. *The Ethiopian famine and relief measures: an analysis and evaluation*. A study prepared for UNICEF in Addis Ababa. Oxford: Balliol College.

Lamb, H. H. 1977. Some comments on the drought in recent years in the Sahel-Ethiopian zone of north Africa. In Dalby, David, Church, R. J. H. and Bezzaz, Fatima (eds) *Drought In Africa*. African Environment Special Report 6. London: International African Institute, pp. 33–38.

Lancaster, Carol J. 1990. The Horn of Africa. In Lake, A. (ed.) *After the Wars*. New Brunswick, NJ: Transaction Publishers/Overseas Development Council, pp. 169–190.

Laughlin, Charles D., Jr. 1974. Deprivation and reciprocity. *Man* (n.s.) **9**, 380–396.

Lemma, Hailu 1985. The politics of famine in Ethiopia. *Review of African Political Economy* (August) **33**, 44–58.

Lirenso, Alemayehu 1983. State policies in production, marketing, and pricing of foodgrains: The case of Ethiopia. *Africa Development*, **8**(1), 72–85.

Lirenso, Alemayehu 1987. Agricultural pricing and marketing policy of Ethiopia: A synopsis. Institute of Development Research, Addis Ababa University, Addis Ababa. Mimeo.

Lirenso, Alemayehu 1990. Villagization: Policies and prospects. In Pausewang, Siegfried, Cheru, Fantu, Brüne, Stefan and Chole, Eshetu (eds) *Ethiopia: rural development options*. London: Zed Books, pp. 135–144.

Locke, C. G. and Ahmadi-Esfahani, F. Z. 1993. Famine analysis: a study of entitlement in Sudan, 1984–1985. *Economic Development and Cultural Change*, **41**(2), 363–376.

Luling, Virginia 1987. Resettlement, villagization, and the Ethiopian peoples. *Development*, **1**, 32–35.

Lycett, Andrew 1992. Agriculture minister calls for better conservation. *African Business* **March**, 16.

Mace, Ruth 1990. Pastoral herd compositions in unpredictable environments: a comparison of model predictions and data from camel-keeping groups. *Agricultural Systems*, **33**, 1–11.

Magrath, John 1991. When farmers take the brakes off. *International Agricultural Development*, **11**(2), 15–16.

Manyazewal, Mekonnen 1992. The economic policy of Ethiopia: Implications for famine prevention. In Webb, Patrick (ed.) *Famine and drought mitigation in Ethiopia in the 1990s*, Tesfaye Zegeye, and Rajul Pandya-Lorch, 19–36. Famine and Food Policy Discussion Paper No. 7. Washington, DC: International Food Policy Research Institute.

Matlon, Peter 1987. The West African semi-arid tropics. In Mellor, J. W., Delgado, C. L. and Blackie, M. J. (eds) *Accelerating food production in sub-Saharan Africa*. Baltimore, MD: Johns Hopkins University Press for the International Food Policy Research Institute.

Maxwell, Simon 1978. *Food aid, food for work, and public works*. IDS Discussion Paper 127. Brighton: Institute of Development Studies.

Maxwell, Simon 1989. *Food insecurity in North Sudan*. IDS Discussion Paper 262. Brighton: Institute of Development Studies.

Maxwell, Simon and Frankenberger, Timothy 1992. *Household food security: concepts, indicators, measurements: a technical review*. Rome: International Fund for Agricultural Development/United Nations Children's Fund.

McCann, James C. 1985. Social impact report: OXFAM/ILCA Oxen–Seed Project. Confidential report. OXFAM America, Addis Ababa. Mimeo.

McCann, James C. 1987a. *From poverty to famine in northeast Ethiopia: a rural history*. Philadelphia, PA: University of Pennsylvania Press.

McCann, James C. 1987b. The social impact of drought in Ethiopia: Oxen, households, and some implications for rehabilitation. In Glantz, M. (ed.) *Drought and hunger in Africa*. Cambridge: Cambridge University Press, pp. 245–267.

Mellor, J. W., Delgado, C. L. and Blackie, M. J. 1987. *Accelerating food production in sub-Saharan Africa*. Baltimore, MD: Johns Hopkins University Press for the International Food Policy Research Institute.

Mercer, A. 1992. Mortality and morbidity in refugee camps in eastern Sudan: 1985–90. *Disasters*, **16**(1), 28–42.

Mirotchie, Mesfin and Taylor, D. B. 1993. Resource allocation and productivity of cereal state farms in Ethiopia. *Agricultural Economics*, **8**, 187–197.

Monod, Theodore 1975. *Pastoralism in Tropical Africa*. London: Oxford University Press/International African Institute.

Moris, Jon R. 1988. *Interventions for African pastoral development under adverse production trends*. ALPAN Network Paper No. 16. Addis Ababa: International Livestock Centre for Africa.

Mortimore, Michael 1988. Desertification and resilience in semi-arid West Africa. *Geography*, **73**(1), 61–64.

Mulhoff, Ellen 1988. *Collection of food and nutrition data and its use in Ethiopia, 1979–1988: A review.* Report to Food and Agriculture Organization of the United Nations. Addis Ababa: FAO.

NCHS (National Center for Health Statistics) 1977. *NCHS growth curves for children: birth–18 years.* Vital and Health Statistics Series 11, No. 165. Washington, D.C.

Neumann, C., Trotle, R., Baksh, M., Ngare, D. and Bwibo, N. 1989. Household response to the impact of drought in Kenya. *Food and Nutrition Bulletin*, **11**(2), 21–33.

Niger (Government of) 1991. *Annuaire Statistique; Series Longues.* Direction de la Statistique et de la Demographie. Ministere du Plan. Niamey, Niger.

NORAD (Norwegian Agency for International Development) 1984. *Report on NORAD-supported health services through Norwegian Lutheran Mission in Ethiopia.* Addis Ababa.

Nyoni, N. 1993. Cattle industry faces challenge. *Development Dialogue (SADDC)*, **3**(8), 9.

OECD (Organization for Economic Cooperation and Development) 1988. *The Sahel facing the future.* Paris.

Oehmke, J. F. and Crawford, E. W. 1993. The impact of agricultural technology in sub-Saharan Africa: A synthesis of symposium findings. Draft report to U.S. Agency for International Development, Washington, DC. Mimeo.

OFDA (Office of U.S. Foreign Disaster Assistance) 1991a. *Famine mitigation: proceedings of workshops held in Tucson, Arizona, May 20–23, 1991, and Berkeley Springs, West Virginia, July 31–August 2, 1991.* Compiled by the office of Arid Land Studies. Tucson, AR: University of Arizona.

OFDA (Office of U.S. Foreign Disaster Assistance) 1991b. Vulnerability assessment. Office of U.S. Foreign Disaster Assistance, Washington, DC. Mimeo.

af Ornas, Anders H. 1990. Town-based pastoralism in Eastern Africa. In Baker, Jonathan (ed.) *Small town Africa: studies in rural-urban interaction.* Seminar Proceedings 23. Uppsala: Scandinavian Institute of African Studies.

Osmani, S. R. 1991. Comments on Alex de Waal's 'Reassessment of entitlement theory in the light of recent famines in Africa.' *Development and Change*, **22**, 587–596.

OXFAM 1984. *Lessons to be learned: drought and famine in Ethiopia.* Oxford: Oxford University Press.

Pankhurst, Alula 1985. Social consequences of drought and famine: An anthropological approach to selected African case studies. PhD diss., University of Manchester, UK.

Pankhurst, Richard 1966. The great Ethiopian famine of 1888–1892: a new assessment. *Journal of History and Medicine*, **21**, 95–124.

Pankhurst, Richard 1984. *The history of famines and epidemics prior to the twentieth century.* Addis Ababa: Relief and Rehabilitation Commission.

Pausewang, Siegfried, Cheru, Fantu, Brüne, Stefan and Chole, Eshetu (eds) 1990. *Ethiopia: rural development options.* Atlantic Highlands, NJ: Zed Books.

Penrose, Angela 1987. Before and after. In Jansson, Kurt, Harris, M. and Penrose, M. (eds) *The Ethiopian famine.* London: Zed Press, pp. 79–102.

Pinstrup-Anderson, Per 1993a. *The political economy of food and nutrition policies.* Baltimore, MD: Johns Hopkins University Press for the International Food Policy Research Institute.

Pinstrup-Anderson, Per 1993b. The food situation in sub-Saharan Africa and priorities for food policy research and donor assistance. International Food Policy Research Institute, Washington, DC. Mimeo.

Rahmato, Dessalegn 1987. *Famine and survival strategies: a case study from northeast Ethiopia.* Food and Famine Monograph No. 1. Addis Ababa: Institute of Development Research.

Rahmato, Dessalegn 1988. *Peasant survival strategies.* Geneva: International Institute for Relief and Development/Food for the Hungry International.

Rahmato, Dessalegn 1990a. A resource flow systems analysis of rural Bolosso (Wolayta). A report. Addis Ababa University, Addis Ababa. Mimeo.

Rahmato, Dessalegn 1990b. Cooperatives, state farms, and smallholder production. In Pausewang, Siegfried, Cheru, Fantu, Brüne, Stefan and Chole, Eshetu (eds) *Ethiopia: Rural development options.* Atlantic Highlands, NJ: Zed Books, pp. 100–110.

Ravallion, Martin 1987. Markets and famines. Oxford: Clarendon Press.

Reardon, R., Delgado, C. and Matlon, P. 1992. Determinants and effects of income diversification amongst farm households in Burkina Faso. *Journal of Development Studies,* **24**, 365–377.

Redd Barna 1989. Bolosso Woreda Project Area (P.4004). Project Background Report. Redd Barna-Ethiopia, Addis Ababa. Mimeo.

de Regt, W. 1993. Can we hope to reunite orphans with their families? *CRDA News* (Christian Relief and Development Association), **4**(3), 4–5.

Reutlinger, Shlomo 1988. Income-augmenting interventions and food self-sufficiency for enhancing food consumption among the poor. In Pinstrup-Anderson, Per (ed.) *Food subsidies in developing countries: costs, benefits, and policy options.* Baltimore, MD: Johns Hopkins University Press for the International Food Policy Research Institute.

Richburg, Keith B. 1992. Ethiopia, Eritrea try free-market capitalism. *The Washington Post,* **February 25**, A13.

Riddell, Roger and Robinson, Mark 1992. *The impact of NGO poverty alleviation projects: results of the case study evaluations.* Working paper No. 68. London: Overseas Development Institute.

Rivers, J. P. W. 1988. The nutritional biology of famine. In Harrison, G. A. (ed.) *Famine.* Oxford: Oxford University Press.

Rivers, J. P. W., Holt, J. F. J., Seaman, J. A. and Bowden, M. R. 1976. Lessons for epidemiology from the Ethiopian famines. *Annales de la Société Belge de Medicine Tropique,* **56** (4/5), 345–357.

Robinson, S. 1989. In Downing, Thomas E., Gitu, K. and Kamau, C. (eds) *Coping with drought in Kenya: national and local strategies.* Boulder, CO: Lynne Rienner.

RRC (Relief and Rehabilitation Commission) 1974a. *Drought and rehabilitation in Wollo and Tigre.* Report of a survey and project preparation mission, October–November, Addis Ababa.

RRC (Relief and Rehabilitation Commission) 1974b. *Hararghe under drought: a survey of the effects of drought upon human nutrition in Hararghe Province.* Addis Ababa.

RRC (Relief and Rehabilitation Commission) 1985a. *The challenges of drought: Ethiopia's decade of struggle in relief and rehabilitation.* Addis Ababa.

RRC (Relief and Rehabilitation Commission) 1985b. *Ethiopia: Review of drought relief and rehabilitation activities for the period December 1984–August 1985 and 1986 assistance requirements.* Addis Ababa.

RRC (Relief and Rehabilitation Commission) 1990a. *Food supply of the crop dependent population in 1990.* Early warning system Meher (main) crop season synoptic report. Addis Ababa.

RRC (Relief and Rehabilitation Commission) 1990b. Summary of NGO activities in Ethiopia, 1989/90. Preliminary report. Addis Ababa. Mimeo.

RRC (Relief and Rehabilitation Commission) 1991. Emergency code for Ethiopia. Draft. Addis Ababa. Mimeo.

RRC (Relief and Rehabilitation Commission) 1992. *Food supply prospects in 1993 (crop growers and nomads).* Early Warning System Report. Addis Ababa.

Russell, N. C. and Dowswell, C. R. (eds) 1993. *Policy options for agricultural development in sub-Saharan Africa.* Mexico City, Mexico: CASIN/SAA/Global 2000.

Ruttan, V. W. (ed.) 1993. *Why food aid?* Baltimore, MD: Johns Hopkins University Press.

Sandford, Stephen 1983. *Management of pastoral development in the Third World.* Chichester: John Wiley for the Overseas Development Institute.

Schellinski, K. 1986. 15,000 orphans in Ethiopia. Briefing Paper. United Nations Childrens Fund, Addis Ababa. Mimeo.

Schneider, S. H. 1977. *The genesis strategy: climate and global survival.* New York: Delta Publishing.

Selinus, Ruth 1971. Dietary studies in Ethiopia: dietary pattern among the Rift Valley Arsi Galla. *American Journal of Clinical Nutrition,* **24**, 365–377.

Sen, Amartya 1981. *Poverty and famines: an essay on entitlement and deprivation.* Oxford: Oxford University Press.

Shepherd, Jack 1975. *The politics of starvation.* New York: Carnegie Endowment for International Peace.

Shipton, Parker 1990. African famines and food security: Anthropological perspectives. Harvard Institute of International Development, Harvard University, Boston. Mimeo.

Shoham, J. and Borton, J. 1989. Targeting emergency food aid: Methods used by NGOs during the response to the African food crisis of 1983–86. Report of a joint study by the Relief and Development Institute and the Human Nutrition Unit, London School of Hygiene and Tropical Medicine. Relief and Development Institute, London.

Sivard, R. 1991. *World military and social expenditures.* Washington, DC: World Priorities.

Sollod, Albert E. 1990. Rainfall, biomass, and the pastoral economy of Niger: Assessing the impact of drought. *Journal of Arid Environments,* **18**, 97–107.

Stahl, Michael 1990. *Constraints to environmental rehabilitation through people's participation in the northern Ethiopian Highlands.* Discussion Paper 13. Geneva: United Nations Research Institute for Social Development.

Stevens, Christopher 1979. *Food aid and the developing world: four African case studies.* New York: St. Martin's Press.

Stewart, Frances 1993. War and underdevelopment: Can economic analysis help reduce the costs? *Journal of International Development,* **5**(4), 357–380.

Stone, Jeffrey C. (ed.) 1991. *Pastoral economies in Africa and long-term responses to drought.* Proceedings of a colloquium, 9–10 April 1990, University of Aberdeen, Aberdeen.

Svedberg, Peter 1991. *Poverty and undernutrition in sub-Saharan Africa: theory, evidence, policy.* Monograph No. 19. Institute for International Economic Studies. Stockholm: Stockholm University.

Swift, Jeremy 1988. *Major issues in pastoral development with special emphasis on selected African countries.* Rome: Food and Agriculture Organization of the United Nations.

Swift, Jeremy 1989. Why are rural people vulnerable to famine? *IDS Bulletin,* **20**(2), 8–15.

Swift, Jeremy 1993. Understanding and preventing famine and famine mortality. *IDS Bulletin,* **24**(4), 1–16.

Swift, J. and Maliki, A. 1984. *A cooperative development experiment among nomadic herders in Niger.* ODI Pastoral Development Network Paper No. 18c. London: Overseas Development Institute. Tareke, Girma 1991. *Ethiopia: power and protest.* African Studies Series No. 71. Cambridge: Cambridge University Press.

Tato, Kebede 1989. Rehabilitation of forest, grazing, and agricultural lands. Technical notes on forestry activities. Food and Agriculture Organization of the United nations, Rome. Mimeo.

Taube, A. 1976. Kejsaruniversitetet och den etiopiska revolutionen. Upsalla University, Upsalla. Mimeo.

Teka, Tegene 1984. Agrarian transformation: The Ethiopian experience. Food and Agriculture Organization of the United Nations, Rome. Mimeo.

Teklu, Tesfaye, von Braun, Joachim and Zaki, Elsayed 1991. *Drought and famine relationships in Sudan: policy implications.* Research Report 88. Washington, DC: International Food Policy Research Institute.

Thebaud, Brigitte 1988. *Elevage et developpement au Niger: quel avenir pour les eleveurs du Sahel.* Geneva: International Labour Organization.

Thesinger, W. 1987. *The life of my choice.* London: Fontana.

Tilahun, Negussie 1984. *Household economics study in Borana and estimation of expenditure elasticities from a sample of Borana households.* Joint Ethiopian pastoral systems study. Research Report 15. Addis Ababa: International Livestock Centre for Africa.

Tiruneh, Andargatchew 1993. *The Ethiopian revolution 1974–1987: A transformation from an aristocratic to a totalitarian autocracy.* Cambridge: Cambridge University Press.

Torry, William I. 1977. Labour requirements among the Gabbra. *Paper presented at the International Livestock Centre for Africa Conference on Pastoralism in Kenya*, Nairobi, Kenya, 22–26 August.

Torry, William I. 1984. Social science research on famine: A critical evaluation. *Human Ecology*, **12**(3), 227–252.

Torry, William I. 1988. Famine early warning systems: The need for an anthropological dimension. *Human Organization*, **47**(3), 273–281.

Toulmin, Camilla 1983. *Herders and farmers or farmer-herders and herder-farmers?* Pastoral Network Paper 15d. London: Overseas Development Institute.

Toulmin, Camilla 1985. *Livestock losses and post-drought rehabilitation in sub-Saharan Africa.* LPU Working Paper No. 3. Addis Ababa: International Livestock Centre for Africa.

Traore, Amadou 1989. Interview with James Ingram, Executive Director of World Food Programme. *The Courier*, November–December.

Turnbull, Colin 1972. *The mountain people.* New York: Simon and Schuster.

Turton, D. 1985. Mursi response to drought: some lessons for relief and rehabilitation. *African Affairs*, **84**(3), 331–346.

UNDP (United Nations Development Programme) 1993. *Human development report 1993.* Oxford: Oxford University Press for the United nations Development Programme.

UNDP/RRC (United Nations Development Programme/Relief and Rehabilitation Commission) 1984. *The nomadic areas of Ethiopia: Part III—The socioeconomic aspects.* Addis Ababa.

UNECA (United Nations Economic Cooperation for Africa) 1985. *Comprehensive policies and programs for livestock development in Africa: problems, constraints, and necessary future action.* ALPAN Network Paper 5. Addis Ababa: International Livestock Center for Africa.

UNEPPG (United Nations Emergency Preparedness and Planning Group) 1987. Food for Work: WFP signs a US$76.1 million project for rehabilitation of forest, grazing, and agricultural lands in Ethiopia. *UNEPPG Newsletter*, July.

UNEPPG (United Nations Emergency Preparedness and Planning Group) 1989a. Food aid assistance to Ethiopia. Addis Ababa. Mimeo.

UNEPPG (United Nations Emergency Preparedness and Planning Group) 1989b. Summary of 1985–1987 emergency operations. UNEPPG Briefing Paper. Addis Ababa. Mimeo.

UNEPPG (United Nations Emergency Preparedness and Planning Group) 1989c. WTOE moves its millionth ton of relief grain in Ethiopia. UNEPPG Briefing Paper. Addis Ababa. Mimeo.

UNEPPG (United Nations Emergency Preparedness and Planning Group) 1989d. Summary of 1988 emergency relief operations in Ethiopia. UNEPPG Briefing Paper. Addis Ababa. Mimeo.

UNEPPG (United Nations Emergency Preparedness and Planning Group) 1989e. A review of UNICEF's relief and rehabilitation strategies and approaches in Ethiopia since 1985. UNEPPG Newsletter July/August: 11. Mimeo.

UNEPPG (United Nations Emergency Preparedness and Planning Group) 1990. Briefing notes on United Nations emergency relief and preparedness activities in Ethiopia. A report prepared on the occasion of the visit by Mrs. Anne Clwyd. Addis Ababa. Mimeo.

UNHCR (Office of the United Nations High Commissioner for Refugees) 1988. *Ethiopia: health and nutrition assessment of southern Sudanese refugee camps in Kefa, Illubabor, and Wolega Awrajas.* Technical Support Service Mission Report 15/88, March 8–22. Geneva.

UNICEF (United Nations Children's Fund) 1988. Quick assessment: Cash-for-food in Ethiopia. UNICEF/RRC emergency intervention evaluation. Draft report. UNICEF/RRC, Addis Ababa.

UNSO (United Nations Sudano-Sahelian Office) 1991. *Reforestation: the Ethiopian experience, 1984–1989.* New York.

USAID (United States Agency for International Development) 1987. Final disaster report. The Ethiopian drought/famine. Fiscal years 1985 and 1986. USAID/American Embassy, Addis Ababa.

USAID (United States Agency for International Development) 1993. Back to the future: Concept paper. Addis Ababa, Ethiopia. Mimeo.

Vallee, Michel 1989. Assessment of WFP assistance in Hararghe region. A report. World Food Programme, Addis Ababa.

Vaughan, M. 1987. *The story of an African famine: gender and famine in Twentieth Century Malawi.* Cambridge: Cambridge University Press.

Vosti, Stephen 1992. Constraints to improved food security: linkages among agriculture, environment, and poverty. In Webb, Patrick, Zegeye, Tesfaye and Pandya-Lorch, Rajul (eds) *Famine and drought mitigation in Ethiopia in the 1990s.* Famine and Food Policy Discussion Paper 7. Washington, DC: International Food Policy Research Institute, pp. 99–122.

de Waal, A. 1987. Famine that kills: Darfur 1984–1985. Save the Children Fund (UK). London. Mimeo.

de Waal, A. 1988. Famine early warning systems and the use of socioeconomic data. *Disasters,* **12**(1), 81–91.

de Waal, A. 1990. A reassessment of entitlement theory in the light of recent famines in Africa. *Development and Change,* **21**(3), 469–490.

de Waal, A. 1991. *Evil days: 30 years of war and famine in Ethiopia.* Washington, DC: Human Rights Watch.

Walters, H. 1989. Agriculture in Ethiopia—The banks' strategy. A draft report of the World Bank. World Bank, Washington, D.C.

Watkins, S. C. and Menken, J. 1985. Famines in historical perspective. *Population and Development Review,* **11**(4), 647–75.

Watts, Jan. 1988. Regional patterns of cereal production and consumption. In Zein, Zein Ahmed and Kloos, Helmut (eds) *The ecology of health and disease in Ethiopia*, Addis Ababa: Ministry of Health, pp. 94–135.

Watts, Michael J. and Hans G. Bohle 1992. The space of vulnerability: structure of hunger and famine. Department of Geography, University of California (Berkeley), Berkeley, CA. Mimeo.

Webb, Patrick 1989a. Baseline survey of Doma Peasant Association. United Nations Childrens Fund/International Food Policy Research Institute, Washington, DC. Mimeo.

Webb, Patrick 1989b. *Baseline survey of Korodegagga Peasant Association*. Report to UNICEF. International Food Policy Research Institute, Washington, D.C. Mimeo.

Webb, Patrick 1992. Food security through employment in the Sahel: Labor-intensive programs in Niger. International Food Policy Research Institute, Washington, DC. Mimeo.

Webb, Patrick, von Braun, Joachim and Yohannes, Yisehac 1992. *Famine in Ethiopia: policy implications of coping failure at national and household levels*. Research Report 92. Washington, D.C.: International Food Policy Research Institute.

Webb, Patrick, and Moyo, Sam 1992. Food security through employment in southern Africa: Labor-intensive programs in Zimbabwe. International Food Policy Research Institute, Washington, DC. Mimeo.

Webb, Patrick and Reardon, Tom 1992. Drought impact and household response in East and West Africa. *Quarterly Journal of International Agriculture*, **31**(3), 230–246.

Webb, Patrick, Richardson, Edgar, Seyoum, Senait and Yohannes, Yisehac 1994. *Vulnerability Mapping and Geographical Targeting: An Exploratory Methodology Applied to Ethiopia*. Washington, DC: International Food Policy Research Institute. Mimeo.

Webb, Patrick, Zegeye, Tesfaye and Pandya-Lorch, Rajul (eds) 1992. *Famine and drought mitigation in Ethiopia in the 1990s*. Famine and Food Policy Discussion Paper 7. Washington, DC: International Food Policy Research Institute.

WFP (World Food Programme) 1986. Project Ethiopia 2488 (Exp. II). Report No. WFP/CFA:21/14–A. Agenda item 14(a). Presented to the twenty-first session of the Committee on Food Aid, 26 May to 6 June, Rome.

WFP (World Food Programme) 1989. *Mid-term evaluation by a WFP/FAO/ILO/UN mission of Project Ethiopia 2488/(Exp. II): rehabilitation of forest, grazing, and agricultural lands*, vol. 1. Draft report. Addis Ababa.

WFP (World Food Programme) 1990. *Appraisal of Project Ethiopia 2488/(Exp. III): food-assisted land improvement project*. Draft report. Main Report. Addis Ababa.

WFP (World Food Programme) 1991a. Status of 1991 minimum emergency food requirements and availability (as of 15 July 1991). Internal File, FDRE Q91. Addis Ababa.

WFP (World Food Programme) 1991b. Interim evaluation summary report on Project Ethiopia 2488 (Exp. II) 33rd Session of the Committee on Food Aid. Agenda item 3(d). Rome.

WFP (World Food Programme) 1993. *Project Ethiopia 2488 (Exp. 3), Report No. WFP/CFA:36/SCP:11/4–A(ODH)* Add. 1, Agenda item 4(a), presented at the thirty-sixth session of the Committee on Food Aid Policies and Programmes, 25 October–29 October. Rome.

White, Cynthia 1984. *Herd reconstruction: the role of credit among Wodoabe herders in central Niger*. Pastoral Development Network No. 18d. London: Overseas Development Institute.

Wilding, Richard 1985. *The history of pastoralism and emergence of the Borana Oromo*. JEPSS Research Report No. 15. Addis Ababa: International Livestock Centre for Africa.

Winer, Nicholas 1989. Agriculture and food security in Ethiopia. *Disasters*, **13**(1), 1–8.
Wolde-Giorgis, Dawitt 1989. *Red tears: war, famine, and revolution in Ethiopia*. Trenton, NJ: Red Sea Press.
Wolde-Mariam, Mesfin 1984. *Rural vulnerability to famine in Ethiopia, 1958–1977*. Addis Ababa: Vikas Publishing House and Addis Ababa University Press.
Wolde-Mariam, Mesfin 1991. *Suffering under God's environment: a vertical study of the predicament of peasants in north-central Ethiopia*. Berne, Switzerland: African Mountains Association/Geographica Bernensia.
Wolde-Michael, H. 1985. The history of famine in Ethiopia. United Nations Development Programme, Addis Ababa. Mimeo.
Wolkeba, Taffesse 1985. Hydrological and meteorological aspect of natural disaster in Ethiopia. Paper presented at the Disaster Prevention Symposium (Ethiopian Red Cross Society), 4–7 September, Addis Ababa.
Wood, Adrian, and Stahl, Michael 1989. Ethiopia: National conservation strategy. Phase one report, first draft prepared for the International Union for the Conservation of Nature. Addis Ababa. Mimeo.
Wood, Charles A. 1977. A preliminary chronology of Ethiopian droughts. In Dalby, David, Church, R. J. H. and Bezzaz, Fatima (eds) *Drought in Africa*. African Environment Special Report 6. London: International African Institute, pp. 68–73.
World Bank 1993. *Ethiopia: toward poverty alleviation and a social action program*. Agriculture and Environment Operations Division. Eastern Africa Department. Green Cover Report. Washington, DC: World Bank.
World Bank/WFP (World Food Programme) 1991. *Food aid in Africa: an agenda for the 1990s*. A joint study by the World Bank and the World Food Programme. Washington, DC, and Rome.
Wörz Johannes, G. F. 1989. *State farms in Ethiopia*. Discussionsschriften 14. Heidelberg: Lehrstuhl für Internationale Entwicklungs und Agrarpolitik.
Wright, K. 1983. Famine in Tigray: Eyewitness report. Report to the Relief Society for Tigray (REST), Mekelle, Ethiopia. Mimeo.
Yitbarek, Kelemu 1988. General distribution. In *Report on Relief Workshop*, edited by the Christian Relief and Development Association, 1–9. Addis Ababa. Mimeo.
Young, Helen 1986. *The evaluation of the OXFAM energy biscuit and other imported foods in selective feeding programs in Ethiopia and E. Sudan*. London: OXFAM.
Young, Helen 1992. *Food scarcity and famine: assessment and response*. OXFAM Practical Health Guide No. 7. Oxford: OXFAM.
Zaman, M. and Parker, B. 1990. Record of disasters in Africa 1980 to present. World Health Organization, Addis Ababa. Mimeo.
Zinyama, L. M., Matiza, T. and Campbell, D. J. 1990. The use of wild foods during periods of food shortage in rural Zimbabwe. *Ecology of Food and Nutrition*, **24**, 251–265.

Index